FUNdamental Mathematics

FUNdamental Mathematics

A Voyage into the Quirky Universe of Maths and Jokes

David Eelbode

ACADEMIA
PRESS

Academia Press
Ampla House
Coupure Rechts 88
9000 Gent
België

www.academiapress.be

Academia Press is a subsidiary of Lannoo Publishers.

ISBN 978 94 014 6261 7
D/2019/45/355
NUR 918

David Eelbode
FUNdamental Mathematics. A Voyage into the Quirky Universe of
Maths and Jokes
Gent, Academia Press, 2019, 311 p.

Tweede druk, 2019

Cover & layout: Flore Swinnen

Foreword

I have no particular talent,
I am only inquisitive.
(Albert Einstein[1])

At some point during our lives, we all feel genuinely misunder-stood. A common fact, almost as universal as the law which says that when you wake up in the middle of the night, wondering whether it's safe to ignore your pressing bladder until the next morning, the answer is always and invariably 'no'. If not during puberty — that time when we all feel misunderstood by our parents and, give or take, the rest of the universe — then at least during excessive Friday night pub crawls, when an unfortunate audience is forced to take our self-proclaimed Words of Genius for what they really are: bags of heavily intoxicated mental rubbish, lining the shadier corridors of our brain like abandoned couches and broken televisions, joylessly waiting to be picked up during the weekly garbage truck procession.

Being a mathematician, feeling misunderstood is part of my repertoire, together with making people at parties feel rather uncomfortable when they ask me what I do for a living, and knowing how much I will have to pay for my groceries, even before the cashier starts sliding my stuff across the miniature party laser light system, marking the end of the conveyor belt.

This is one of the reasons why I decided to write the book you are now holding in your hands: by the time you have reached its final sentences, I hope that you will have gained some insight into the workings of a mathematically inclined mind. Because it may not always be that obvious, but whenever it is not concentrating on a

---

1   Every text needs at least one Einstein quote.

calculation or a proof, nor longing for another cup of coffee, the somewhat overdeveloped mental muscle sitting on top of my body is essentially trying to do what yours is doing: getting a firmer grip on this thing called 'daily life'. In a rather peculiar way though, since it is only fair to say that my bulimic brain has a ravenous appetite for food for thought. Like a baseball coach in his dugout, my brain is ferociously munching its way through the days, masticating even the most mundane of facts, processing them into propositions which can be tested and perfected, trying to translate the results into theorems that can survive the times. Or at least your attention span, as this book contains the fruits of my mental foraging over the last few years: definitions and theorems about things like tofu and polar bears, the mathematician's way of trying to make sense of things.

The second reason why I decided to write this book stems from the fact that I never really understood why the vast majority of people consider my field of expertise, by and large, to be their least favourite subject at school. French and PE can basically do what they want to win this notorious race: at the end of the road, the Queen of the Sciences[2] will still be waiting for them. Freshly showered, casually solving Sudokus while enjoying a colourful cocktail. Deep down this makes me feel inherently sad — like seeing elderly men shuffle through the red light district on Christmas Eve. Or worse still, some people even try to upgrade their cool by admitting that they suck at mathematics — often blaming their high school teachers while they are at it. Something tells me that you cannot accomplish the same result when outing yourself as an illiterate person, right? With this book, therefore, I hope to build a neural connection between what may very well be two completely disconnected regions in your brain: the centrally located Skyscrapers of Fun, and the abandoned Maths Barracks.

---

2   Not my words: it was Carl Friedrich Gauss (1777-1855) who once said that mathematics is the Queen of the Sciences.

In order to situate my third reason to write this book, we need to go back to 2015 — my very own annus horribilis. A year marked by burn-out and depression, two hands around the same throat. After an intense struggle at home — clinging to a sofa and a bottle of pills — I decided to leave Antwerp behind for a few months. It was either Antwerp or my life, so I had nothing to lose. I essentially felt like a reptile at the local zoo: everyone just assumed that I was still alive, because the light in my terrarium was still burning, but what people were really seeing was the faint afterglow of a comet that had plummeted through the atmosphere, crashing into rock bottom, producing a huge crater in which what once was a sparkly body of energy slowly started to decay into a pile of smouldering ashes and dust.

Until I finally found the courage to pull the plug out of my maddening mental fuse box: I decided to take unpaid leave, and went to Tokyo for four months. A surprising choice according to my friends and family, but it made perfect sense to me: Japan is an awe-inspiring universe in itself, with a giant attractor at its center. A pulsating heart which made my blood flow again too. It was in Tokyo — with its slender women who are often as wobbly on their high heels as the skyscrapers during one of the many earthquakes — that I regained my stability. It was there — in the world's capital of sleep deprivation, with its hard-working inhabitants who can doze off more easily on a packed train than I can fall asleep after three sleeping pills, a bottle of wine and a five hour rendition of a melting snowball — that I found the will to get up in the morning again. It was in Japan — where the contrasts are as sharp as the knives with which they fillet their tunas — that I shed my old skin. It was there, miles away from mathematics, that I realised how much I missed her. In a sense, this book thus doubles as a gigantic love letter: the music of reason was my first love, and it will be my last.

Over the past few months, people have often asked me: 'So what kind of book are you writing?' Right from the start I knew that I wanted to write the type of book that ends up on a coffee table.

Or on that little shelf in the toilet where most people keep a few outdated magazines and a crossword book. It can definitely be devoured in one go, but I actually recommend you to read a few pages and then leave it there for a while; like most of Tool's albums, olives and facial hair, it has to grow on you.

So is it a mathematical book? Yes and no. It most certainly looks like one — it contains definitions, theorems and formulas — but then again, it is not. I have merely attempted to combine my biggest passions in life (science, language, humour, food, travelling and music) throwing them into a blender, hoping to create something which can both make you laugh and teach you something. About maths, science and the venturings of an inquisitive mathematician in this world.

This book would have stayed an abstract concept forever — just like black hole tourism and me wearing my Isis t-shirts in public again[3] — if it wasn't for a few people who helped me turning it into the actual thing you are now holding in your hands. First of all, I would like to thank Lies (for going along with this crazy plan of mine in the first place), Isaac (for your valuable advice throughout the publication process) and Flore (who must have cursed me on more than one occasion for my constant nagging about ill-placed spacing and ugly fractions). Also Tamsin and Wouter deserve a special mention here: the latter is a good reader of proofs, the former is an outstanding proofreader.

Next, I would like to thank a few people who make my life on Planet Maths even better than it already was with maths only: Alberto (my one and only Klimentska-brother), Paul and Rudi (as prime as colleagues can possibly be), Annick and her posse, Mariska and Suzy, Greet and Christine (Eager Scientists for the win), Werner and Stijn, Bart (I promise not to mention

---

3   Isis used to be this utterly brilliant metal band, before the name was hijacked by malevolent people.

'Gobelijn' here), Matthias plus the Twin-Tim (if you ever start a band together, promise me to call it 'the Higher Spin Sons'), the Cliffordians at Ghent University (count Vladimir included) and my students (giving proper meaning to my life as a mathematician).

Finally, a big shout out to the Earthlings who are always eagerly awaiting me when I return from yet another trip in that vast space commonly referred to as 'my mind'. My parents (for allowing me to study maths in the first place), *watashi no nihonjin no kazoku* (see you at Ware), Wibbie and Macky (let that be a lesson, thou shalt not contract names), Nick and Ilse (owners of the only chihuahua that will ever matter), Elke and Sarah (VdM = Very dedicated Mates), HGP and the BLB-crew (I bet half of this book was conceived at your place), Evy (curiosity may have killed the cat, it did give birth to something better), Sofie and Tuur (I do hope you realise I will start reading this book out loud when you ask me to babysit), Laurent and Mathieu (<3) and Popol (luring me into the lovely arms of the Queen of the Sciences).

And you, of course, for buying (or any other verb, for that matter) this book and giving it a chance. Have fun along the way!

David Eelbode

Warning: this book contains functional nudity, an absolutely enormously huge amount of exaggerations, n mathematical symbols (with $n \in \mathbb{Z}^+$), fucking profanities, mathematical puns, footnotes, traces of peanuts,[4] dead puppies, a completely irrelevant amount of tofu, definitions (these are statements that explain the meaning of a word or phrase), ambiguous alliterations, mental mathematics, plenty of coffee, jokes which some scientists may find a little offensive,[5] small additions for math geeks only (where 'small' can be any $\epsilon > 0$), a few wombats and the absolute minimum of formulas.[6]

---

4   The sum over all peanuts on the diagonal.

5   I did try to make them easy enough, so that also the engineers can have a laugh.

6   The theoretical proof for this was eliminated by the editor though, as it involved an argument containing a lot of mathematical equations.

Contents

1

A Space Odyssey

Space. It seems to go on and on forever.
Then you get to the end, and a monkey
starts throwing barrels at you.
(Philip J. Fry, *Futurama*)

You should really read this chapter if ...

- you don't really know what 'space' actually is.

- you have always wondered why you can't divide by bean curd.

- you don't believe that headbanging is encoded in the laws of nature.

- you don't know the link between coconuts and our planet's poles.

1.1 Space on stage

We all need some space. A big personal bubble around us. Far away places we can travel to, in order to develop our sense of time. Room, for improvement. Nooks and crannies, to cram away our junk. Territory, to fight over. The occasional piece of space in cake — or, if you are an astronaut celebrating your birthday at work, the other way round. Unknown lands, to explore. White space between letters, so we can take a breath and give meaning to words — why else did they make that bar the largest key on a laptop? Dark space to stare into on a starry summer night, so we can lie on our back and let the quiet void fire our philosophical thought generator.

'Why are we here?'

'Are we alone here, or is there something out there?'

'Does a block of tofu really exist or is it a soy-based illusion?'

Definition 1.

Tofu: *one of the biggest remaining mysteries in the field of Contemporary Culinary Philosophy, next to famous problems such as 'Is a cheese cake tart or pie?', 'Is it morally acceptable to put pineapple on a pizza?' and 'Is the hot dog a sandwich, or not?'*[7]

Not only philosophers, but also exact scientists are desperately trying to find out what tofu is. Based on the fact that nothing ever seems to change when you add it to a dish — tofu has neither taste, nor smell — mathematicians are claiming that tofu is the neutral element in the group of food additives. Put differently: it is the edible version of the number zero (you may remember from your maths classes that adding zero to something never changes anything, unless we are talking about school reports). Seeing tofu as a kind of zero also explains why you get total nonsense when you try to divide by it.

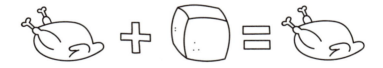

BASIC EXAMPLE OF TOFU ARITHMETIC

Theoretical physicists have taken a keen interest in tofu because it seems to be the only substance in the visible universe which does not contain traces of peanuts. This has led them to believe

---

7   America's National Hot Dog and Sausage Council (NHDSC) ruled that they are not. As the organisation's president put it: *Limiting the hot dog's significance by saying it is 'just a sandwich' is like calling the Dalai Lama 'just a guy'.*

that bean curd is the absolute culinary vacuum: a zero-calorie state, which can only be obtained by extracting all the smell and flavour from matter. The only problem is that this does not explain the abundance of meat in the cosmos, which is why some people have conjectured the existence of dark tofu, the non-vegan counterpart of regular tofu. So far it has not yet been observed, but NASA (the National Aroma and Scent Administration) has plans to launch a space probe in 2022. This would be a natural follow-up to their most recent project: sending the Marslander Space-Eyed to our planetary neighbour, in order to see whether there once was whisky on the red rocks.

As to those other big questions of what space actually is, and where it came from, a mathematician already uses a completely different notion than, say, a philosopher or a physicist. For the latter, space is essentially a stage on which things exist, events happen and history unfolds. And just like love for furry purring kittens, it seems boundless, intangible and ubiquitous.[8] I often find myself wandering around this imaginary stage as if I were an extra in a movie without a script — utterly clueless and hoping for a free lunch — so this image does correspond with my gut feeling: space is a plane of existence, but then in three dimensions instead of two. However, mathematicians do not feel satisfied with this description. Like doing a job interview in a tracksuit: it may sound comfortable, but it is just not formal enough. What they want is a sound definition for space, an issue that I will try to address in a later chapter ('Setting Up Space').

Note that when physicists say three dimensions (backwards and forwards, right and left, up and down), they actually mean four. But this requires the inclusion of time as a fourth dimension — from past through present to future — which is precisely what Albert Einstein suggested when he stupefied the world with his theory of relativity, in the early 20th century. There are even people who claim that we live in a universe which has as many as

---

8   See also: the internet

ten or eleven dimensions — at least six of which are required in order to properly understand tofu, and one clogged up with lost socks. Some of these people spend their lives on a psychiatric ward wearing straitjackets, because they are said to be delusional. Others are even less fortunate and spend their lives as a string theorist.

Definition 2.

String theory: *an advanced mathematical framework in theoretical physics which attempts to address some of the most fundamental questions in black hole physics and early universe cosmology. It is seen as a serious candidate for the so-called Theory of Everything, a mathematical model describing all the fundamental forces of nature (gravity, electromagnetism, the weak nuclear force, the strong nuclear force and Jedi mind power) and all forms of matter. Until the day string theorists succeed in explaining why we sometimes have to rotate a USB-stick twice before it fits in the port, it can safely be ignored.*[9]

---

9   The mathematically schooled reader may see this as a proof that USB-sticks are spinors.

Definition 3.

Dimensions: *it is not easy to explain in full detail what a dimension exactly is, but luckily enough there is a nice parallel with the number of people having sex together. When there is just one person involved (one dimension), the only option is to move back and forth. Despite being a hands-on introduction to the world of dimensions, this is definitely not the most interesting case as it does not leave room for intersections. The next case (two dimensions) is definitely the most familiar one: it is taught at school and lends itself perfectly to visualisations on a screen, which may explain the many movies and websites covering this case. The situation in three dimensions is the most interesting one, as it allows manoeuvring around in a variety of ways which are inconceivable in the previous cases (left or right, in or out, top or bottom). Anything from four dimensions onwards makes most people feel rather uncomfortable, although some do prefer ten or even eleven (the swing theorists).*

Apart from the number of dimensions it has, the precise origin of space is also still a topic of heated debate amongst cosmologists and theoretical physicists. It actually divides them into competing camps quarrelling over, well, nothing really: one of the prevailing conjectures is that our universe — or even a multitude of parallel universes — came into existence out of absolute nothingness. Not unlike that adamant pimple that often sets up camp overnight, right in the middle of my forehead, where I somehow seem to lack the appropriate amount of skin to squeeze. There is probably even a parallel universe out there in which none other than Shakespeare fathered modern cosmology:

From zero to here? Oh!
A causative quantum coincidence.
Colliding clouds of quarks.
Coalescing chunks of matter.

Ecce,
the quirky creation of this celestial clusterfuck,
collectively referred to as our cosmos,
in a coconut shell.
To be from not to be,
that was the question.

(from 'Much Ado About Nothing')

All that is, was and will be, a universe much too big to see,[10] generated by some sort of abstract glitch: apart from a few religious
fanatics — who seem to have taken this whole Big Bang Theory to
a very unfortunate different level — this is of course where most
of the religions and their followers pray to differ, as they all seem
to have some sort of creation myth to adhere to. During my own
struggle with a course on quantum field theory at university, I
often asked myself: who needs physics when you can have something as simple as a cosmic egg, hatched by a celestial bird? Or an
ancient Hindu story, according to which every life form on Earth
originates from clarified butter — their take on *Ghee-nesis*, as
worthy of spreading a Holy Word as can be.

Definition 4.

Quantum field theory: *this is another advanced framework in*
theoretical physics, aimed at explaining the behaviour of subatomic
particles through a combination of special relativity and quantum
mechanics. Despite being one of the most succesful frameworks
ever devised by physicists, emerging from the work of generations
of geniuses throughout the 20th century, quantum field theory itself
lacks a rigorous mathematical foundation. So this means that from
a mathematical point of view, quantum field theory is like holding

---

10   Oh, did I already mention that you get extra points for spotting references to
      lyrics?

the remote control[11] *over your head with a stretched arm when you want to switch channels but the batteries are running low, while pressing the buttons twice as hard as normal: it somehow works, but nobody knows why.*

I am going on a bit of a sidetrack here, but my favourite creation myth is definitely this story about the Japanese god Izanagi, who pushed his 'jewel encrusted spear' into 'the primal ooze of our planet' and, when pulling out, 'spilled a salty substance' that created the island of Onogoro.

He then took his soon-to-be wife Izanami to this huge stretch of dried ejaculate, where they married and had sex. After that, his spouse gave birth to more islands, and several gods — one pair for each substance that escaped from her body (vomit, urine and faeces). So according to this extract from Japanese Genejizz it all started with a terrible toilet trinity on Spunk Island. Who would have thought that the first commercial pitch for an MTV adult dating show is more than 1,000 years old?

Not everyone agrees with this opposition between science and religious traditions though. Some scholars have opted for a different tactic, as they are desperately trying to reconcile the current theories in physics with theology. They believe that string theory actually leaves room for a prime instigator. 'God is not a DJ,' they claim. 'God plays the guitar and He struck the very first chord.' The part about God not being a DJ sounds perfectly plausible to me: there is no Bible verse which says 'In the beginning there was absolutely nothing, and then He pressed the play button', and yet we have David Guetta. But if He did have a guitar, then

---

11   I don't know whether you ever noticed this, but most remote controls have a numerical pad with a tiny raised dot in the middle (right at the fifth button). I find this a bit strange: if you cannot see the number 5 button, you probably shouldn't be watching television in the first place.

it must have been a heavily distorted one in drop D,[12] as it surely caused an astrophysical mosh pit, and it seems to have drenched the cosmos in feedback (described by Pythagoras as 'the music of the spheres', and now referred to as 'the cosmic background noise').

You have to admit, with its hard rocks, heavy metals and all that black and dark stuff out there in space, it looks like our universe does fancy the kind of music people usually associate with long hair, a beer belly and the satanic horn sign. The stuff you won't typically hear at televised wedding ceremonies (although I suppose the Iron Lady must have been an Iron Maiden at some point, so that is at least one opportunity missed). I personally find this prime instigator scenario as hard to digest as a brick wrapped in cellulose. That being said, we've only just started our story so it is probably smarter at this point to steer away from the murky waters of dispute and controversy. So let us treat this topic like a beached jellyfish: better left untouched.

12   Drop D tuning is an alternative way of guitar tuning, frequently used in heavy metal and its various subgenres. There is a technical explanation for it, but it essentially amounts to tuning the lowest string of a guitar in such a way that plucking it reduces even the most severe forms of constipation to mere bowel dust.

Definition 5.

Mosh pit: *a gathering place for people participating in a frenetic communal dance, often referred to as 'moshing' in scientific circles. Although nowadays mostly observed at metal and punk rock concerts, it was accidentally invented by a linguist translating the idiom 'when push comes to shove' from English into body language. Moshing is an extremely difficult dance style to master, because it requires the performers to converge towards a thermodynamical equilibrium that allows the spontaneous formation of circular vortices known as 'circle pits'. This was proved (it really was, this is not a joke) by Jesse Silverberg and three of his colleagues at Cornell University in 2013, and published as a genuine research article. They found that moshing people can recreate speeds that have the same statistical distribution as the speeds of particles in a gas. Note that there also seems to exist a slightly less violent two-dimensional version of moshing, better known as 'breakdancing'. The underlying mathematical principles of this dance style are still unknown at the time of writing.*[13]

The Maxwell-Boltzmann distribution: you probably bumped into — and possibly tripped on — the phrase 'the same statistical distribution as the speeds of particles in a gas' in the previous definition. This phrase actually refers to a formula (yet another opportunity to trip) for a certain quantity $f(v)$ which was thoroughly studied by physicists. In all its glory, this quantity looks as follows:

$$f(v) = \sqrt{\left(\frac{m}{2\pi kT}\right)^3} 4\pi v^2 \exp\left(-\frac{mv^2}{2kT}\right) \;.$$

---

13   This is actually a general rule in some mathematical disciplines: contrary to what you may expect, there are problems that tend to become *easier* when you consider them in *more* dimensions. Unfortunately, as I will explain in a later chapter, remembering where you last put your keys is not one of them.

'This scares me, do we really need it?'

Please bear with me for a second, I promise you a striking conclusion in return. The formula for $f(v)$ above is a nice example of a so-called *distribution function*. It was named after two famous 19th-century scientists (Ludwig Boltzmann, an Austrian physicist who made essential contributions to the theory of heat and entropy, and the Scotsman James Clerk Maxwell, best known for the mathematical description of electromagnetism), and it literally describes the distribution of the speeds v of the particles in a gas (or, surprisingly enough, people attending a metal concert). The thing is that not everyone moves at the same speed in the mosh pit, and what this formula does is to quantify this behaviour.

This works as follows: the mathematical expression for $f(v)$ should be seen as some sort of abstract machine. Given the temperature T at the concert venue and the mass m of the moshing people, you can plug in an arbitrary speed v, and after some calculations – hereby making use of the formula for $f(v)$ – the machine will spew out the probability $P(v)$ that someone in that group will be moving at that very speed.

$$\boxed{\textbf{INPUT: speed } v} \xrightarrow{\;f\;} \boxed{\textbf{OUTPUT: probability } P(v)}$$

Let us consider a simple example. If you plug in an input speed v equal to zero, this machine will tell you how likely it is that someone is *not* moving (that's what zero speed does to you). Now for $v = 0$, you can see that the output probability becomes zero too.

'Wait, what?'

> Indeed, not only does the formula for $f(v)$ contain a factor v^2 – which means that it becomes zero when the speed is equal to zero – also the probability $P(v)$ will be zero for $v = 0$. So what the formula tells us, is that there is a zero probability that someone will be 'moving' at speed $v = 0$. Put differently, this essentially means that no one will be standing absolutely still at the concert venue. Which should not really come as a surprise: we were talking about metal and punk rock, remember? So this is just a mathematical proof for the phenomenon called 'headbanging'.

1.2 On coconuts and spherical cows

Now going back to our original question of what 'space' actually is then, is there a way to see this concept through mathematical glasses? Obviously yes, otherwise these first few pages would have been a dramatic buildup towards the biggest anticlimax since the Millenium bug (often referred to as the Year 2000 problem[14]). Mathematicians usually define a space as 'a set, to which some sort of extra structure is added'. If you cannot wrap your head around this sentence yet, don't worry: over the course of the following chapters, I will explain what a 'set' is and how we can add more 'structure' to it — turning a mere set into a space ('Setting Up Space'). Along the way, we will bump into a few mathematical problems that can be related to the properties of space: we will stuff things into it, slice it up, tile it and even get lost in its corners ('Space To Slice And Stack'). Not only those corners: we will occasionally find ourselves lost in what I personally find one of the scariest spaces I know of.

'Seriously? A whole section on Ibiza?'

---

14   If you do not know what this is, no worries. It probably means that you were born after 2000. Erhm, sorry, I mean 1900.

No, I am talking about the mathematician's very own world: I will introduce you to the peculiar language he uses ('Space Between Letters'), I will reveal what happens in those secret meetings the uninitiated sometimes refer to as 'a mathematical conference' and I will sketch her professional interests and activities ('Space To Explore').

Speaking about problems relating to space and its properties, I would like to end this introductory chapter with two deceptively easy questions about our planet Earth, that one place in space we are all too familiar with.

Definition 6.

Earth: *a remarkably inappropriate name for a planet that is — statistically speaking — just Ocean.*

Problem 1. Suppose you are somewhere on Earth, and you decide to make a trip: first you go south for 1,000 kilometres, then you go east for 1,000 kilometres, and finally you go back north for 1,000 kilometres. If you now end up where you started, *where exactly did you start*?

Problem 2. If you travel north long enough, you will end up travelling southward at some point. However, if you travel east, then you will always be travelling eastward, and never westward. *Why is that?*

Martin Gardner (1914-2010)

Answer 1. This is actually a famous problem from Martin Gardner, one of the most colourful characters in mathematics in my opinion. He sadly passed away in 2010, but he will always be remembered as the father of recreational mathematics: apart from the columns he wrote for *'Scientific American'*, he also wrote several dozen books, full of truly beautiful mathematical puzzles and problems aimed at a general audience — a lot of space related to problems, that is.

The first puzzle in his book *'My Best Mathematical and Logic Puzzles'* is called 'the Returning Explorer': *an explorer walks one mile due south, turns and walks one mile due east, turns again and walks one mile due north. He finds himself back where he started and shoots a bear. What colour was the bear?* Formulated like this, chances are that you see why 'the North Pole' is the correct answer to this puzzle.[15] Well, I say 'the' correct answer, but the most surprising thing about this puzzle is that there are *infinitely many* correct answers — unless you really insist on shooting that bear. So if you did not think of the North Pole as a solution, you've got a second chance here: can you guess what other places on Earth also solve this puzzle?

As a matter of fact, it gets even better: there is an infinite number of infinitely many correct answers. This is genuine mathematics: when solving the problem is not enough, you can turn counting the solutions into a new problem. In the third chapter ('Setting Up Space'), we will even see how hard this new problem of counting the infinite number of solutions becomes, as it is related to one of the biggest revolutions in the history of mathematics. How does *that* score on the Cliffhanger Scale?

---

15 Bonus question: what colour was the explorer's jacket? For an answer to this question, see a not-so-famous theorem about polar bears (in chapter 6, 'Space To Explore').

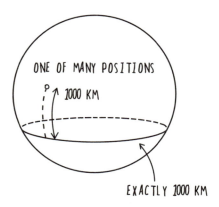

Answer 2. One can argue that this apparent difference between the combos North-South and East-West is the result of a simple convention: fixing the Earth's poles. The geographic poles, I should add, as these are actually quite different from the Earth's magnetic poles. The former can be seen (well, not literally) as the unique points where all the meridians meet, the latter are a manifestation of a physical phenomenon. Note that the Earth's magnetic poles seem to be on the move. Although this is bad news, it does have the advantage that you can use this as an excuse next time you are late. 'Oh, sorry, I thought you meant relative to the *magnetic* poles!'

Apart from this more conventional explanation, there is also a surprising connection with a famous theorem: the 'Hairy Ball Theorem' in topology, a mathematical discipline we will come back to in chapter 6. Needless to say that adding the word 'theorem' when you type the name of this famous result into the Google search bar is strongly recommended. If you do forget to add it, you may — amongst others — bump into the name of a former Forth Wayne mayor: Harry Baals. Just in case you think he had the most unfortunate name in human history: Mike Litoris and Willie Stroker are genuine names too.

The Hairy Ball theorem answers a question which sounds like the start of a bad joke: what do you get when you lock a mathematician in a room with a comb, a pencil and a coconut?[16] The thing is: you cannot comb a hairy ball flat without creating a cowlick, and there is a theoretical proof for this fact. As is often the case in my field of expertise, it sounds slightly less romantic if you use proper maths speak (as in 'there are no non-vanishing continuous tangent vector fields on even-dimensional n-spheres'), but this is what the theorem essentially amounts to. These spheres in higher dimension will be introduced in chapter 8, when we have a look at hyperballs, but you do know the easiest examples already: the 1-sphere is just a circle, and the 2-sphere is what soccer players pass around. What this theorem then says, in great generality, is that something fancy *can* be done with hairy circles ($n = 1$), but *not* with hairy balls ($n = 2$). If this kind of distinction between even and odd cases surprises you, just think of the last time you tried to sleep on n ears, where n is 1 or 2.

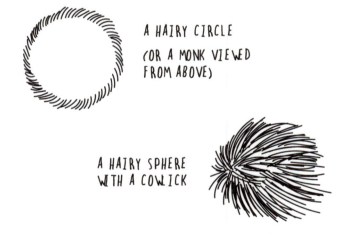

A HAIRY CIRCLE
(OR A MONK VIEWED FROM ABOVE)

A HAIRY SPHERE
WITH A COWLICK

A well-coiffed circle (n = 1) versus a cowlicked sphere (n = 2)

---

16 Not to be confused with the answer to the question: what do you get when you lock a mathematician in a room with a monkey and a coconut? In that case, you end up with Martin Gardner's favourite puzzle, 'the monkey and the coconuts' (see e.g. Wikipedia).

The Dairy Ball theorem: amongst theoretical physicists, the phrase 'a spherical cow' is a well-known humorous metaphor for highly simplified scientific models for complex real life phenomena. Theoretical physicists will often reduce a problem to the simplest form they can imagine, to get more grip on the calculations, but these simplifications always hinder the model's application to reality. So whereas scientists would love to solve their problems under the most general conditions, this is not possible in reality because this requires taking too many variables into account, and this then makes the equations too hard to solve — even by computers. I mean, predicting where a Tesla Roadster launched into space will *exactly* be in 1,000 years from now is a very hard problem, unless you do the calculations for a spherical car in a vacuum. In hindsight, I think it is a real shame that people tend to use coconuts to illustrate the Hairy Ball Theorem. Just for once, starting from a spherical cow would have been a perfect metaphor.

I have noticed that in many books, authors tend to include a list of recommended literature at the end of a chapter. As this book is all about trying to tingle non-conventional senses — and since I have devoured hours of music during the writing of it — I have opted for a list of recommended tracks. After all, people often say that music is pure mathematics. I wish you at least one musical discovery.

Recommended listening for this chapter

Artist	Song title
Black Box Revelation	Where Has All This Mess Begun
The Cinematic Orchestra	Time and Space
Iron Maiden	Caught Somewhere in Time
Monster Magnet	Space Lord
65daysofstatic	No Man's Sky
The Prodigy	Out of Space
Envy	Ticking Time and String
Faithless	God Is a DJ
Isis (the band)	All Out of Time, All Into Space
Metallica	Through the Never

2

Space Between Letters

> *The proper definition of a man*
> *is an animal that writes letters.*
>
> (Lewis Carroll[17])

You should really read this chapter if ...

- you want to see the mathematical formula for a penis.

- you want to know what peas have to do with the world hunger problem.

- you are convinced the only thing that can't be measured is the weight of your dream.

- you feel like testing your 'counting up to 1,000' skills.

2.1 ABC or 123?

Some people say that mathematics is a language of its own. I do not fully agree with this: languages are still means to communicate, so as long as I cannot use them to order a decent bowl of Japanese noodles in miso broth with freshly chopped onion, a seasoned soft-boiled egg and pinkish pickled ginger on the side, they do not qualify as such in my opinion. However, I do recognise that some knowledge of algebra allows you to convey a few essential messages.

---

17   Known by most people as the author of *Alice's Adventures in Wonderland,* but what most people don't know is that he was a gifted mathematician as well. For instance, he worked in linear algebra, which is the mathematical study of vector spaces — one of the most essential mathematical spaces out there.

For instance, next time you are out on a date you may consider whispering $(x^2 + y^2 - 1)^3 = x^2y^3$ into your dating partner's ear, because the connoisseurs know that this equation represents a heart.

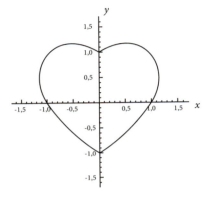

The heart as the graph of the equation $(x^2 + y^2 - 1)^3 = x^2y^3$.

You can also skip the subtleties and just go for the function below. Probably about as close as you can get to crossing a maths formula with a dick pic.[18]

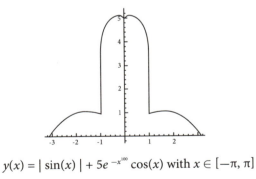

$$y(x) = |\sin(x)| + 5e^{-x^{100}}\cos(x) \text{ with } x \in [-\pi, \pi]$$

---

18   Knowing the workings of the internet and its inhabitants, I am really looking forward to opening my inbox after the publication of this book. Or maybe not.

Graphs and equations: scientific books often contain plenty of graphs — this book forming no exception — so let me explain how these visual representations for mathematical equations come into being. If you put a dot in a grid on all the positions (x, y) for which a certain equation is satisfied, you end up with a collection of dots which mathematicians refer to as 'the graph of that equation'. You can think of this graph as a massive abstract spreadsheet, where the columns are labelled by names of women (the XX-axis), rows with names of men (the XY-axis), and where a cell is coloured if the couple (f, m) satisfies a certain criterion (like 'm knows the brother of f' or 'f crossed the street to avoid m').

The equation $y = x^2$, for instance, can be represented by the parabola below (its graph), because every point on that curve labelled by a pair of coordinates (x, y) has the property that the y-coordinate is equal to the square of the x-coordinate. For example, the points $(2, 4)$ and $(-3, 9)$ belong to the graph, because $2^2 = 4$ and $(-3)^2 = 9$.

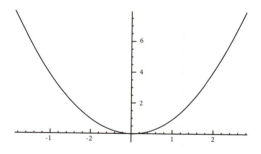

The parabola $y = x^2$.

The equation $(x^2 + y^2 - 1)^3 = x^2y^3$ looks more complicated, but the idea behind the graph is the same (I actually think this is a pretty accurate description of doing mathematics: start from a simple notion, and stretch it until it becomes scary and complicated). The equation is satisfied when we choose $x = 1$ and $y = 1$, because $(1^2 + 1^2 - 1)^3 = 1^21^3$, which means that the point in position $(1, 1)$ will belong to the graph. On the other hand, the equation is not satisfied when we choose $x = 1$ and $y = 2$, as $(1^2 + 2^2 - 1)^3 \neq 1^22^3$, so this point does not belong to the graph.

The upshot is that if you draw *all* the points (x, y) for which the equation $(x^2 + y^2 - 1)^3 = x^2y^3$ holds, you will end up with a heart-shaped graph. Or, given the fact that this may very well be your first attempt ever at drawing a curve defined by a polynomial equation of degree 6, at least a graph into which you have put your own heart.

Regardless of whether mathematics is indeed a proper language or not, it is definitely true that mathematicians make extensive use of the alphabet: interesting objects and special numbers are given a name — they do not call it a brainchild for nothing I guess — which often consists of a single letter. Some of these are universal, like the letter π for what I like to call 'the mathematical Starbucks number'

$$\pi \approx 3.14159265358979323846264338327950288419716 9 \ldots$$

as it seems to pop up everywhere you look, whereas others may depend on the context or even the flavour of the author. Even the least commonly used letters in the English alphabet play a role: the letter z is typically used for complex numbers, q usually denotes a quaternion or a quantum deformation parameter (quirky stuff, *quoi*), j often stands for a summation index[19] and x is of course best known as the unknown variable — doubling as a stereotypical representative of maths phobia.

Definition 7.

Maths phobia: *according to cognitive psychologists and neuro-scientists, it is a genuine feeling of tension and fear that interferes with maths performance, or exposure to formulas and mathematical descriptions in general. This was concluded after an elaborate study in which a collection of 131,071 data points (which happens to be a Mersenne prime number) in a five-dimensional space was submitted to a product-moment correlation test, hereby using a radially symmetric correlation function which satisfies a partial differential equation of order 4 over an algebraically closed field. If you are starting to feel slightly sweaty and nervous having read that passage, I am very sorry. I merely wanted to explain something by providing enough details. If you think that was a cruel joke, think of how the people suffering from hippopotomonstrosesquipedalio-phobia must feel: by the time the doctor has explained to them that it means 'fear of long words', they are probably already having a full-blown attack. And unlike people suffering from maths phobia, they cannot be helped with prescription pharmaceuticals.*

For those readers who just don't believe that maths phobia is a real thing, let me tell you about this computer science student I once taught calculus to. He approached me after a lecture, red-faced and profusely sweating. 'Sir,' he said, 'I'd like to attend your classes this semester, but I have difficulties concentrating and I believe this is due to maths phobia panic attacks.' When I asked him for more details, I got a foam-mouthed adolescent, breaking into a 17-minute rant about his former high school maths teachers, involving so many profanities that it felt like listening to an

---

19 Chances are that you are now thinking: 'What?! They told me that the letter *j* stands for a complex unit!' This probably means that you are an electrical engineer.

audio version of a swear word dictionary. Now I am obviously not a trained psychologist — not even an untrained one — but I remembered this documentary about voodoo I once saw on the television, and I decided to give it a try. So I told this guy: 'Listen, I really need you to let go of your former math teachers. I therefore suggest the following: if you still happen to have the books they were using, I want you to burn them and bury the ashes in your backyard.' One week later, this student was back, and (somewhat to my surprise) he looked a completely different person. So after my lecture, I asked him whether my little plan had worked. 'Thank you so much, sir,' he replied. 'I did exactly like you told me: I burnt them all, buried the ashes and did a little dance afterwards. That felt so good. But I did forget to ask you one thing last week: what was I supposed to do with the books?'

Speaking about maths books and letters, mathematicians use so many different letters that the standard Roman alphabet is often too limited: just open the nearest maths textbook — it might be on that pile where your copy of *The Illustrated History of Embroidery in Kyrgyzstan* is gathering more dust than a mysophobic's wipe cloth — and chances are that you will come across Greek characters, gothic letters and a bunch of symbols which look like they were found in a Dingbat catalogue. I suppose most people are familiar with the four basic operations $(+, -, \div, \times)$ and the occasional integral sign $\int$ (the devil's signature), but even I have to blink twice when I come across one of the following symbols:

$$\curlywedge \quad \Cup \quad \oslash \quad \bowtie \quad \multimap \quad \sqcap \quad \doteqdot \quad \oint$$

I got them from a maths textbook myself, but I have absolutely no idea what they mean. If I really have to guess, I would say that the sequence above forms a genuine sentence in an alien dialect from a different solar system, and that it stands for 'always watch out for low flying pianos'.

Symbols like these make me think of a particular lecture I once attended, held during the first Summer School I ever participated. It was delivered by this British professor who had clearly prepared his presentation using a version of Windows which was not compatible with the one installed on the local machines: when he double-clicked his file, it became clear that all his mathematical symbols had been replaced by genuine Zapf Dingbat symbols. So while the audience was desperately trying not to crack up with laughter, this man was pointing at formulas containing card suits, crayons, envelopes and snowflakes. Something along the lines of

$$\tfrac{1}{\maltese} \int_0^{\pi} \sin(2x + \circledast)d\boxtimes = \frac{\sqrt[\heartsuit]{\heartsuit\spadesuit+\clubsuit}}{2} \cos(\maltese x^2) + \varreturn$$

The funny thing about this was not that it actually worked — once you have been told that the telephone symbol stands for a projection operator it all makes sense, it is just a matter of convention after all — but that the speaker was able to maintain his composure, staring into this sea of chuckling faces. I cannot remember what the man lectured about, but I do remember that he essentially proved that there is no competition for British upper lips on the Universal Stiffness Scale — not even from wedding shoelaces and Japanese business etiquette rules.

This imminent lack of letters which can be used in mathematics is one of the main reasons why I tend to agree with people who say that it makes sense to study Chinese nowadays. Not just because it might be useful from a more economical point of view, but it could also be an inexhaustible source of mathematical symbols. Then again, although this connection between Chinese and maths will come quite naturally for some people — the ones who know that *chǎomiàn*, or stir-fried noodles, equals 17 — I am afraid that once we start calculating with Chinese characters, it will become all Greek to most people.

Definition 8.

Chinese: *an artificial language, devised by a team of linguists and Western tattoo artists in 1951, after an incident involving a man who wanted to get a Helvetica 'friendship' tattoo on his forearm but ended up with a Tahoma 'firendship' (sic).*

As irrelevant as it may sound, during my career as a researcher I already found myself engaged in more than one heated discussion with colleagues about the choice of a letter or a name for a new mathematical object. Not because we enjoy being nitpicky — I could now add that this is not completely true and that some of us actually do, but this probably makes me come across as one of them — but because bad notations defy the purpose, which is the condensation of information into a simple string of representative symbols. So choosing a notation which confuses people is like using greyscale posters for a Skittles advert: it just does not make sense. Not only in mathematics can a badly chosen name lead to confusion. I mean, just think of what could have happened if Jessica Biel (married to Justin Timberlake) had named her son Batmo.

2.2 Mathematical world records

The idea of using shorthand symbols, the modern mathematical notation, was not invented until the 16th century. Before that, things were mostly written out in full sentences, apart from numerals and a few other basic symbols. For instance, Fermat's celebrated last theorem originally read as follows:

Cubum autem in duos cubos, aut quadrato-quadratum in duos quadrato-quadratos, et generaliter nullam in infinitum ultra quadratum potestatem in duos eiusdem nominis fas est dividere.

This very same message can also be written as follows:

$$(\forall n \in \mathbb{N})\big(n > 2 \Rightarrow \neg\big(\exists(a,b,c) \in \mathbb{N}_0^3 : a^n + b^n = c^n\big)\big) \ .$$

You don't even have to be familiar with the precise meaning of these symbols to see that this formulation is shorter, hence more convenient, and universal. Now I fully understand that to some people, symbolic statements like this seem scary. I have the same feeling towards purple-haired grannies — the type of women who often carry around a dog whose name is longer than the creature itself, and who draw on their own eyebrows, sometimes so badly that they seem to wander around in a state of constant amazement (these women scare the bejesus out of me). Through repeated exposure, most mortals seem to succeed in overcoming the former — fear of the arcane flair of formulas — but I still seem to struggle with the latter.

Fermat's last theorem: the symbolic statement above says something about the numbers we all learnt in kindergarten, when maths was all about solving food-related issues such as 'if I have five apples, three olives and 19 tomatoes, how many pieces of fruit do I then have in total?' The kind of numbers you can count on your fingers, so to speak: numbers like 1, 17, 42 and so on. Mathematicians call these numbers the 'natural numbers' or the 'positive integers' (we will meet them again in the chapter 'Setting Up Space', when the set $\mathbb{N}$ is defined).

Fermat's theorem tells us that amongst all the positive integers, the numbers $n = 1$ and $n = 2$ play a very special role: these are the *only* values for which one can find three (positive) integers a, b and c such that $a^n + b^n = c^n$ (all different from zero of course, otherwise it is not that difficult). For $n = 1$, this is trivial, assuming you still remember that the first power of a number is just that number. You can *choose* any two integers a and b, there will always be an integer c such that $a^1 + b^1 = c^1$. This is called 'adding numbers'. For $n = 2$, it is still possible to find three integers for which the relation is satisfied, but you can no longer choose the numbers (a, b, c) *arbitrarily*.

We obviously still have that $1^2 + 1^2 = 2$ — so far, so good — but the problem is that 2 should be equal to c^2 for some integer c, and this is not possible. Don't get me wrong: the square root of 2 does exist, but it is not an integer (it is a decimal number, to say the least). On the other hand, we have $3^2 + 4^2 = 25 = 5^2$ and $7^2 + 24^2 = 625 = 25^2$, which means that $(a, b, c) = (3, 4, 5)$ and $(a, b, c) = (7, 24, 25)$ are examples of triplets of integers for which $a^2 + b^2 = c^2$. We call these solutions 'Pythagorean triples', as they can be seen as the lengths of the sides of a rectangular triangle, which satisfies Pythagoras' theorem (see also the picture below). Some historians of mathematics claim that this very theorem was already known to the Babylonians, centuries before Pythagoras was born, but no one seemed to have bothered telling him, making this the oldest problem in non-communicative geometry.

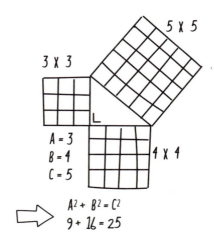

The Pythagorean triple $(3, 4, 5)$

However, for some strange reason the numbers $n = 1$ and $n = 2$ seem to be an exception: take any other integer number n and you will never find natural triples (a, b, c) for which the relation $a^n + b^n = c^n$ holds. Not because a paranoid president fired them under suspicious circumstances, or because mathematicians have not been trying hard enough to find them: they simply do not exist. Let me put it this way: it may not be clear what Pythagoras himself has put into his theorem, but Fermat's theorem is essentially saying that he definitely got the most out of it.

It is probably worth pointing out again that Fermat's last theorem only applies to these positive integer numbers, the simplest amongst many different kinds of numbers. If you are allowed to use non-integer numbers, there are lots of triples (a, b, c) for which the formula holds — even for n bigger than 2 — but then you need the kind of numbers you cannot typically count on your fingers. These are called 'irrational numbers', elements of the set $\mathbb{R}$ (which we will also encounter in the chapter 'Setting Up Space'). A number like $\sqrt[3]{2} \approx 1.25992105$ for instance, which allows us to write down the relation

$$1^3 + 1^3 = 2 = (\sqrt[3]{2})^3 .$$

This clearly is a Fermat-like statement for n equal to 3, but it involves a non-integer number. Note the use of the symbol '$\approx$' in the example: in mathematics, this universal bacon sign stands for 'approximately equal to'. This may look like scientific laziness, as maths claims to be an *exact* science, but it is nothing less than an identity crisis: irrational numbers have a decimal expansion that goes on forever, and never repeats itself — I wish the same thing could be said about some television series. So if it weren't for the fact that we can refer to the number above as 'the cube root of 2', it would essentially be impossible to describe it in full detail: $\sqrt[3]{2}$ is characterised by — and counting on your hands thus requires — an *infinite* number of digits. 'Are you sure there is no one out there with an infinitely many fingers, who could get this job done?'

As a matter of fact, I am not, but it is safe to assume that if there indeed were a creature of that sort, the adjudicator would still be busy counting.

Definition 9.

Adjudicator: *since it is these people's responsibility to critically evaluate lottery drawings and Guinness World Record attempts, I think it is only fair to say that they are better at counting than some of the mathematicians. Well, better than me at least, because I only started paying attention in maths class when the numbers became letters: for me, the 'abc' does not refer to the alphabet, but to a non-commutative product of three algebraic generators. This is why I actually preferred reading stories to counting and doing mental arithmetic as a school kid. I cannot recall a single maths book title from that time, but I do remember enjoying* The 117 Dalmations *and* Snow White and The Nine Dwarfs.

Adjudicators may be good at counting, but they are definitely surpassed by some of the people who have a Guinness World Record involving mathematics to their name. People like Rajveer Meena,[20] recalling the first 70,000 digits of π in ten hours (that is two digits per second on average — a phrase which also seems to apply to some soccer players' bank balances) or Vikas Sharma, calculating 15 large number square roots in one minute.

This mental circus act is obviously quite an astonishing feat, but as with that more conventional form of juggling: I am afraid that I need to see something which requires more balls to be really impressed. Or hyperballs, that would be really impressive.

---

20   During the preparation of this book, I learnt that Akira Haraguchi had already recited 100,000 digits in 2006, but the Guinness Book of World Records did not validate this record. Whether this now makes him eligible for the world record 'reciting the most number of π-digits before not being granted a world record' is still a topic of heavy debate amongst contemporary philosophers.

Definition 10.

Guinness Book of World Records: *a ridiculously lengthy list of established world records, from genuinely funny (the fastest 100 metre hurdle wearing swim fins, 14.82 seconds), through fairly far-fetched (the most sexual innuendos in a potato-themed rap song, 35 at the time of writing) to downright dull (the slowest time to close a drawer, currently standing at 17 minutes and 21 seconds: I have watched this video at 128 times its original speed and it still seemed less exciting than the average panda's sex life). But just like Fermat triples of integers (a, b, c) with n bigger than 2, some Guinness World Records will never be established because they are mathematically impossible: unless there exists at least one adjudicator with insomnia, we will never know how many times a trained herd of sheep can jump over a fence.*

...41, 42, 43, 44,...

Fermat's last theorem forms a nice exception, in that it effectively gained its entry in the Guinness Book of World Records as 'the longest-standing maths problem'. It was conjectured by the French mathematician Pierre de Fermat (1607-1665), who claimed to have an ingenious proof 'which was too large to fit the margin' — a common trick which is known in academic circles as 'a proof by lack of space' (there you go, another advantage of having access to higher dimensions: way more margins). Most experts say that it is highly unlikely that he really had a proof though.

It was eventually proved by Andrew Wiles in 1995, a whopping 385 years after it was formulated (some whiles can take a long time), but it will always be known as 'Fermat's Last Theorem'. This is the mathematical equivalent of what happened to the new wave classic 'Tainted Love' by Soft Cell, the English synthpop duo, as only a few people seem to know that this song was written by Ed Cobb and originally recorded by the American singer Gloria Jones in 1964. This kind of process even has its own name: 'Stigler's Law of Eponomy' states that no discovery is ever named after its original discoverer. Sadly enough, it was coined by Stephen Stigler himself in 1980, which gives this law the same credibility as a fashion blog sponsored by Crocs.

Note that Fermat's last theorem does something rather peculiar: it tells us with an *absolute* certainty that something does not exist (combinations of integers *a, b* and *c* for which a certain mathematical relation holds). My mother never achieved this back when I was a kid: no matter how hard she tried to convince me that there were no such things as ghosts, the little thinker trapped inside my body had already established that alleged non-existence is hard to verify in real life. Truth be told, I never found a beast under my bed — nor my closet nor my head — but that was far from ruling out their existence. This is due to a peculiar asymmetry: in order to demonstrate that something exists, it suffices to find at least one example. But you can never be sure that something does not exist, until you have left no stone unturned. Well, unless you want to prove the non-existence of unturned stones — good luck with that.

Mathematics sets itself apart in this regard: non-existence results occur quite often (the Hairy Ball Theorem from the previous chapter provides a nice example). They even play an important role, as they prevent researchers from studying irrelevant or plainly impossible things. I know what you are thinking right now: 'Isn't *all* research in pure maths irrelevant?' This question will be addressed in more detail in the chapter 'Space To Explore', but let me at least say this: although the connection between mathematics and reality is not always that obvious — saying that this is an understatement is an understatement — one could argue that in the end, its purpose is to pave that bumpy road running between abstract models and real-world phenomena. Well, I call it a bumpy road, but I do mean that steep, bouldered mountain path, hewn out of pure granite by an army of tenured thinkers, marked by the blood, sweat and tears of traumatised students, crossing the Forest of Frustration and the Desert of Disillusioned Dropouts, overlooking the Vast Oceans of Unwanted Minus Signs and the Great Plains of Pi's Decimals, cautiously curling around the City of Corrupted Conjectures and Cursed Counterexamples.

DESERT OF
DISILLUSIONED
DROPOUTS

FOREST OF
FRUSTRATION

VAST OCEANS OF
UNWANTED MINUS SIGNS

GREAT PLAINS OF
PI'S DECIMALS

CITY OF CORRUPTED
CONJECTURES AND CURSED
COUNTEREXAMPLES

Applied mathematics obviously does that paving by default. But what some people tend to forget — sadly enough, including the ones approving fundamental research grant proposals — is that pure maths also serves a purpose in this respect: it is a natural breeding ground for many of the tools which are used by applied mathematicians. Some of them were even introduced and studied by theoreticians in times when the application itself would have been called witchcraft. Just imagine the Ancient Greek Eratosthenes (276-194 BC), asking the Cyrene Maths Society to fund his interest in prime numbers, 'because at some point in the far future these will enable people to send their safely encrypted credit card details over a global wireless network, so that they can buy a kilogram of cheese from their ergonomic office chair by clicking a mouse'. They would not understand him of course: the kilogram was not invented until the late 18th century.

Definition 11.

Prime number: *an integer number greater than 1 which can only be divided by 1 and itself. Despite there being an infinite number of primes, as shown by Euclid, finding them is not an easy task. It was Eratosthenes who proposed a simple algorithm for finding prime numbers. His method is now known as the Sieve of Eratosthenes, the second most famous household utensil in science after Occam's Razor. However, we have long entered the realm where his method is no longer adequate and more advanced techniques are thus required. This makes prime numbers to mathematicians what dinosaur bones are to palaeontologists: every once in a while someone finds an even bigger example — an event which then earns the finder a mention on the news (though rarely, pardon the pun, during prime time), which makes a select group of people very envious but makes the majority of people wonder whether there aren't more useful things to do with their tax money.*

Prime numbers are far from the only example of a theoretical subject in maths whose threads were used to weave the nest in which modern-day applications were birthed. For instance, most people struggle with the idea that mathematicians can consider the square root of a negative number, but electrical engineers work with $\sqrt{-1}$ all the time. Granted, in the same way as cleaners work with the Second Law of Thermodynamics[21] — they can safely ignore the theoretical details for all their intents and purposes — but this rather elusive *imaginary* number (the square root of minus one is definitely not a *real* number) does underpin the application. Not only that, it is also a nice example of an abstract concept for which the notation heavily depends on the people using it: mathematicians refer to the square root of minus one as the number i, whereas engineers often call $\sqrt{-1}$ the number j. They then usually add that the letter 'i' is reserved for the electric current, and that they don't want to mistake one for the other. I find this somewhat of a lame excuse, because you can easily tell imaginary numbers and electric currents apart: just touching them with your finger will do, no? Then again, maybe the engineers are merely trying to sympathise with Marvin the Paranoid Android, whose bout of depression got worse when he mistook a mathematical fact for an existential problem: 'How do you mean? I is not real?'

_____

21   Have you ever wondered why throwing all the pieces of a jigsaw puzzle on the floor rarely produces a completed puzzle? Or why a dusted bookshelf doesn't stay clean forever? This is the Second Law of Thermodynamics at work: in nature, chaos prevails.

Developing the tools for practical applications is of course not the only purpose of pure maths — the fundamentalists amongst the fundamental mathematicians would probably go so far as to claim that this has never even been a purpose in the first place. In a sense, mathematics also explores the very capacities of itself, like a doctor performing self-surgery to see what is possible, and what is not. We will encounter a beautiful example of this philosophy, mathematics saying something about itself, when we meet Kurt Gödel (in the next chapter). From this point of view, the afore-mentioned non-existence results are like those strips of tactile warning studs along the highway, startling mathematicians awake whenever their mind seems to go astray. Whenever I see tactile strips, I always like to think that they are the world's best kept secrets, written in braille.

Definition 12.

The Laws of Thermodynamics: *roughly speaking these laws summarise the properties of energy and its transformations from one form into another (given shape by, amongst others, Boltzmann and Maxwell — the physicists we met in the previous chapter). It is formulated in terms of concepts like temperature, thermal equi-librium and entropy — the ever-increasing measure for chaos and disorder which cleaners are trying to get under control. Despite being widely accepted as a successful theory in physics, there are a few things I would like to point out. First of all, an isolated system is supposedly in thermal equilibrium with itself. This means that the overall temperature tends to evolve towards a constant, with heat flowing from warmer regions to colder regions. But how do you then explain a woman going to bed? Because over the years, I have come to the conclusion that no matter how hot they are, they always have cold feet. Secondly, the Laws of Thermodynamics famously forbid a so-called perpetual motion machine. However, suppose you take two Japanese businessmen who are ignorant of each other's rank and you make one of them bow to the other one.*

*As local etiquette rules stipulate that the lowest in rank bows last,
this sets in motion a never-ending chain of nodding and bending —
as environmentally friendly as it can get. How do you explain that?*

Not only the letters that mathematicians use can be different, even
their words often have a distinct meaning and therefore tend to
confuse lay people. Groups are not really what you expect them to
be, and neither are rings or fields. As a matter of fact, but without
going into too much detail here, each of these concepts is defined
as a set with an additional structure. Straight lines are examples
of curves, and bodies of revolution are rarely seen at Rage Against
The Machine concerts, wearing Che Guevara t-shirts. Adding and
subtracting are essentially the same to mathematicians — as with
the zodiac, it is all about the signs — and when they answer 'Yes!'
to the question 'Do you want tea or coffee?', they are not being
(annoyingly) smug but merely answering according to the standard laws of logic.

Definition 13.

Logical disjunction: *in mathematics, the word 'or' refers to 'the
inclusive truth function', which makes statements of the form 'A
or B' true whenever at least one operand (A/B) is true. So logically
speaking, a statement of the form 'I am wearing a metal shirt or I
am eating an apple' is true as soon as at least one of these statements
is correct (but I will definitely not be naked and eating a banana[22]).
This 'inclusive or' should be contrasted with the 'exclusive or' (sometimes symbolised by XOR), which is true if either A or B is true, but
false if both are true. So next time you ask someone 'Do you want
coffee or tea?' and that person answers with a useless 'Yes!', the standard laws of logic allow you to serve a cup containing a coffee and
tea combo (yikes!). If that person starts complaining, you can always
fault him for not specifying he was using the 'exclusive or', rather
than the standard 'inclusive or'.*

---

22   I did warn you about functional nudity at the start, didn't I?

One of the things that I have learnt from past relationships is that both the inclusive and exclusive 'or' should not be confused with the 'or' appearing in seemingly harmless questions from your partner. For example, my girlfriend once said: 'Here's a hypothetical question. Suppose you could spend the night with one of my friends, whom would you choose: Sofie or Sarah?' Being a trained mathematician, I got really excited that evening, because technically speaking this meant that a threesome was a valid logical option. I did decide to play it safe though, and I went for an 'exclusive or' — although I had to resist the urge to point out that this was not specified. Later that night, on the couch, I concluded that there exist questions of the form 'A or B' whose answer is always false, regardless of whether A or B is true. I suggest calling this 'ROR' from now on, for 'relationship or'.

A	B	A OR B	A XOR B	A ROR B
True	True	True	False	False
True	False	True	True	False
False	True	True	True	False
False	False	False	False	False

2.3 It's just words folks, just words

Needless to say that as with most of the ingroups based on profession, mathematicians also have their very own lingo to describe their findings.[23] I will often make use of this jargon throughout this book, so I have decided to spend the rest of this chapter describing some of these words and expressions in more detail.

2.3.1 Axiom

An 'axiom' is to mathematics what a 'dogma' is to religion: it is a set of beliefs which — once settled — is no longer questioned or doubted. That is not to say that some axioms have never been changed or discarded throughout the ages. Consider the picture below, for instance: your mission — should you choose to accept it — is to draw a line going through the dot p, parallel to the given line L. I am pretty sure that everybody ends up drawing the same line, and that is not a coincidence: most of you probably learnt at school that there is only one line meeting the requirements, not the other line (nor, if you thought drawing it would summon a pink unicorn serving you rainbow soup with meatballs, your expectations).

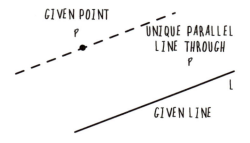

GIVEN POINT
p
UNIQUE PARALLEL LINE THROUGH p
L
GIVEN LINE

---

23   For the mathematicians reading this: if L denotes an arbitrary language and L_μ the set of words used in mathematics, then the probability $P_b(w_1, ..., w_n; L)$ that a normal person who speaks L will not understand you if you make a sentence of b words using the terms $w_1, ..., w_n \in L_\mu$ (with $b \geq n$) is given by the number $(^n/_b)^m$ with $m > 1$ a constant which depends on the language L.

This premise is known as the *Playfair axiom*. It is related to Euclid's fifth axiom, a mathematical postulate that played a crucial role in the history of mathematics.

In all its glory — and complexity — it reads as follows: *if a straight line falling on two straight lines makes the interior angles on the same side less than two right angles, the two straight lines, if produced indefinitely, meet on that side on which the angles are less than the two right angles.*

<gulp>

I decided to include the precise statement, because it may have the same effect on you as it had on most mathematicians: it baffled them, all the way up to the 19th century. For despite the fact that it seems to express an obvious truth — just keep Playfair's formulation in mind — it feels suspiciously more elaborate than Euclid's other axioms (such as 'two points determine a unique line', 'all right angles are equal' and 'never divide by tofu').

This seemed to suggest that the fifth axiom — in sharp contrast to the other axioms — was actually a theorem, and for more than 2,000 years people were therefore trying to *prove* Euclid's fifth axiom, starting from the other axioms.

This is like doing sign language wearing oven gloves: completely pointless. And yet, these pursuits of a proof for the fifth axiom did not come to an end until the 19th century, when it was finally realised that it is *just* an axiom. A rule of play, so to speak, which has to be used as the premise for further reasoning, leading to the type of conclusions mathematicians usually call 'theorems' (see a bit further). But this means that different axioms lead to different theorems. Think of chess, for instance: it is a strategic board game, with its own set of rules (the 'chess axioms'). If we now change one simple rule — after three minutes of chess, the players have to engage in three minutes of boxing, and both sports keep alternating — one gets a completely new game, known as 'chess boxing', with its very own fans and competitors. I actually once tried it myself, and I can confirm that playing chess after boxing is indeed very challenging: I couldn't even *move* my pieces wearing those boxing gloves.

Going back to Euclid's fifth axiom, we can thus expect new kinds of geometry if we replace the original rule (a unique parallel line) with one of the — admittedly, less intuitive — alternatives: there is *no line* through the point which is parallel, or there are *at least two lines* through the dot which are parallel (which then always implies that there are *infinitely many* parallel lines going through the dot). Both scenarios are possible, and merely lead to a different branch of mathematics with its own theorems. The first option, no parallel lines, leads to *spherical geometry* (figure 2 below). I will come back to this case when I talk about pirates and metric spaces in the next chapter. The second option, infinitely many parallel lines, leads to hyperbolic geometry — the geometry of the pringle or the saddle (figure 3 below).

Different kinds of geometry: flat, spherical and hyperbolic.

Definition 14.

Hyperbole geometry: *this is the most awesome branch of mathematics ever invented. It is based on the axiom that given a straight line and a point which does not belong to that line, one can always overdraw a unique line out of proportion which exaggerates the result.*

This particular example involving Euclid's axioms essentially teaches us the same thing as swallowing a handful of cherry tomatoes or three cadmium batteries does: some things are easier to digest than others. The *axiom of empty set* for instance, which essentially says that there exists such a thing as a set that contains nothing, is taken for granted by most mathematicians. Well, most of them: some non-believers claim that when you place an empty set in the middle of a living room, it will soon thereafter contain at least one cat (if you keep wondering what these 'sets' are: they will be the topic of the next chapter).

THE AXIOM OF EMPTY SET

NORMAL VERSION — EMPTY

CAT OWNER VERSION — NOT EMPTY

SCHRÖDINGER'S CAT VERSION — EMPTY & NOT EMPTY

Much more contested is the so-called *axiom of choice,* which does have its opponents. This particular axiom says that if you start from an infinite number of collections of things, each containing at least one object, you can make a new collection of things which contains one object from each of these collections. In other words: you can pick one sock from an infinite number of washing machines, leaving an infinite number of people wondering where that other sock went to (get over it people, life socks). This often makes me wonder why no one has ever come up with the idea of selling socks in packs of three. Sounds like a perfectly legitimate business model to me.

Despite the fact that the previous paragraph contains the word 'infinite', a word which often leads to situations in which people's intuition is fooled (including yours, see next chapter), this axiom of choice does not bother me. I adopt the same philosophy when it comes to philately: as long as I do not have to do the collecting myself, I have absolutely no problem with it.

'But why is it contested then?'

Well, first of all: when the infinity is 'too large', it is absolutely not clear how to collect the socks. It is never really *doable* — on account of it being an infinite amount in the first place — but sometimes simply devising a proper collection strategy becomes a problem, even if it may be a purely theoretical one. And I use the word 'problem' as a euphemism here, because it is simply impossible. So invoking the axiom of choice often leads to a mathematical object which is then proven to exist on theoretical grounds, but which can never be constructed in real life.[24] I started accepting these rather elusive mathematical objects once I bought a cookbook on traditional Japanese cuisine: a nice collection of dishes which are proven to exist — how else can you explain the pictures — but I can never reconstruct them myself. So once again, I have absolutely no problem with this axiom.

The second reason why it is contested is the following: you can use the axiom of choice to prove mathematical results which are — how shall I put this mildly — *completely bonkers*. One of those results is the 'pea and sun paradox', which says the following: it is possible to cut a small ball, say the size of a pea, into as few as five pieces and to reassemble these into a much larger ball, say the size of our sun. And the freaky part is that this process does not involve changing the shape of the pieces. So there is no need for stretching, bending or attaching 3D-printed extensions.

As strange as it may sound, the reassembly process only involves moving around and rotating. Apart from hoping that my dance partner wouldn't notice my eager erection — uncomfortably nesting itself between our swaying bodies — this is also precisely what I did during a slow dance at a party as a 16-year-old: moving around and rotating.

<the sound of a book slammed shut>

---

24 You cannot even write it down in explicit form, which thus makes it the complete opposite of the perfect partner: this concept can be perfectly described on paper, but doesn't exist.

Are you still here? Good, because I swear this is not black sorcery: when it comes to creating something from nothing, some of my ex-girlfriends are still in the running to become the uncontested number one (or am I really the only one who sometimes finds himself in a discussion which seemingly came out of nowhere, cornered by what often turns out to be a circular reasoning to begin with?). Nor is it a mathematically sound solution to the world's hunger problem — not least because parking a pea the size of the sun on our planet sounds like a problem on its own.

As a matter of fact, the 'pea and sun paradox' is not even a paradox! Just like a dusty vacuum cleaner: it does not *sound* right, but there is nothing wrong with it. The 'pea and sun paradox' is a proper mathematical theorem, with a proof that makes perfect sense. Mathematical sense that is, because it is obviously as counterintuitive as the fact that 'flammable' and 'inflammable' are synonyms.

The secret lies in the strange dissection used in the proof of this theorem: the proverbial pea is to be partitioned into five pieces according to a method that cannot be implemented in real life — the pen is mightier than the sword indeed. One has to resort to abstract group theory,[25] as it cannot be done using standard DIY geometry — anything involving a ruler, a compass and a protractor. I guess this goes to show that mathematics has a few things in common with soccer: it is hard to be successfull when you keep hitting the bars, and there are certain things you can do with your head, but not with your hands.

It should not come as a surprise that non-standard cutting techniques lead to non-standard pieces: using this abstract partitioning method involving groups, one arrives at a collection of five 'non-measurable' sets, or — *nomen est omen* — sets which cannot be measured. Abstract reasoning dictates their existence, but it is not possible to assign a real number to them which tells you what their length, area, volume, foot size, body weight or IQ is (our way of measuring something). If you *were* to attach a number to 'its size', you would immediately run into contradictions, which is obviously what mathematicians are trying to avoid at *any* cost — together with scratching their genital areas after having used blackboard chalk sticks.[26]

---

25 A mathematical discipline studying the 'groups' which were mentioned earlier in this chapter. Very broadly speaking, it is the study of symmetries. I think the least thing the inventors could have done when 'group theory' was invented, was to settle for a palindrome as a name for this area of maths.

26 I am talking from experience here: during my first ever lecture at university, I must have repeatedly scratched my balls. After the lecture, a friendly student sheepishly pointed at my jeans, bearing white chalk marks accentuating my private parts. I learnt my lesson, though, that day: when I lecture now, I only use blue chalk.

So the only way out here is simply to conclude that the pieces cannot be measured. And once you can accept this, you can maybe see why there is no paradox involved: the rearranging takes place on the abstract level where your building blocks *have no size*.

So what appears to be black magic — the fact that one can alter the volume of a ball without enlarging the constituents — is essentially done in a framework where words like 'volume' and 'changing the size' are meaningless.

Note that the incongruous 'pea and sun paradox' is the stronger version of what is officially known as the Banach-Tarski paradox (another misnomer, since this is not a paradox either). This theorem, proved by Stefan Banach and Alfred Tarski in 1924, explains how to 'decompose' a ball into as few as five (non-measurable) pieces, which can then be reassembled into two identical copies of the original ball. I can only conclude that Jesus' miracle known as 'The Feeding of the 5,000' is an application of this theorem to five bread rolls and two pufferfish.

> Anagram(mathematician) = thematic mania
>
> Anagram(eleven plus two) = twelve plus one
>
> Anagram(integral) = anger lit
>
> Anagram(Banach-Tarski) = Banach-Tarski Banach-Tarski

Chances are that I lost you two paragraphs ago, because we are not really used to things that have no measure. Since these exotic 'non-measurable sets' lie at the very heart of the Banach-Tarski paradox, I will try to give you an example involving an ordinary circle with radius 1. This is a fairly simple object which has a 'size' – usually referred to as the circumference of that circle – given by 2π. Using the aforementioned non-DIY geometry, I will now show that you can disassemble this circle into a collection of infinitely many, mutually disjoint sets (so that no two sets have points in common) and these sets will all be *non-measurable* – literally having no size.

'Wait, what?! No size, at all?'

Yes, no size whatsoever. I know that this sounds very counter-intuitive, so once again: bear with me. I will explain how this is possible, but then you will have to keep the following things in mind:

(i) The number π is not a rational number, but an *irrational* number (like the number $\sqrt[3]{2}$ we met earlier, when we talked about Fermat's last theorem). This therefore means that π can not be written as a quotient of two positive integers (or a fraction, as we say). `Wait, someone once taught me that π is approximately equal to 22 divided by 7'. That person was definitely not lying, but 'approximately equal' is still not the same as 'equal'. Approximating irrational numbers with fractions is like drawing someone's portrait: no matter how accurate it is, it's just not the real thing. The fact that some numbers can never be expressed by integer fractions was actually discovered in Ancient Greek times. Legend has it that the discovery of the first number which could not be written as a fraction of integers, the square root of 2, led to Pythagoras going a bit irrational himself: he was so angry with Hippasus – who is often credited with the discovery of $\sqrt{2}$ and its irrationality – that he threw him overboard, after which the latter drowned. Safety in numbers, you say?

(ii) In what follows, we will refer to points on the circle as angles expressed in radians. Angles and radians, not angels and radiance. People are used to thinking of angles in terms of degrees (a right angle, for instance, is 90 degrees), but mathematicians usually express them in terms of radians. Now, I've got good news and bad news. The good news is that both systems are related: if you remember that an old-fashioned angle of 360 degrees is equal to 2π radians, then you should be able to handle the rest of this explanation. The bad news is that it's 12 o'clock too: climate change is real. Anyway, if we see this clock as a circle and we start measuring from 12 in clockwise direction, then the numbers 3 (or a 90 degree angle), 6 (or 180 degrees) and 9 (270 degrees) correspond to the angles $\frac{\pi}{2}$, π and $\frac{3\pi}{2}$ when expressed in radians – a classic example of the Rule of Three. Moreover, you can always add 2π (or 360 degrees) to an angle: this will catapult you half a day forward in real life, but on the clock you cannot see this. Mathematicians call this the 'modulo operation', and this thus says that after a full rotation, you are back where you started – well, unless you are a USB-stick. This for instance means that the numbers $\frac{\pi}{6}$ and $\frac{13\pi}{6}$ correspond to the same point on the circle (one o'clock, at an angle of 30 degrees), as their difference is equal to 2π and this can be ignored (the modulo rule).

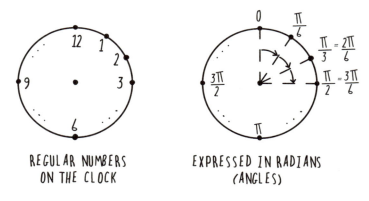

REGULAR NUMBERS ON THE CLOCK

EXPRESSED IN RADIANS (ANGLES)

(iii) There's at least one thing about size that does matter: its *additivity*. Be it length, area or volume: these notions all share the property that the measure of a sum is equal to the sum of the measures. Pour a volume of rum, two volumes of tequila, one volume of wodka, one volume of cointreau, tree volumes of whisky and two volumes of gin together, and you get yourself ten volumes of hangover. It even works if you have infinitely many pieces[27]: the Greek philosopher Zeno of Elea could not get his head around it, but half a cake plus a quarter of a cake,

---

27　For the mathematically inclined reader: yes, countably many. For the other readers: see the next chapter for this subtlety.

followed by one eighth of a cake and so on is simply equal to one piece of cake. Or in numbers: $\frac{1}{2} + \frac{1}{4} + \frac{1}{8} + \frac{1}{16} + \ldots = 1$ (and just like your favourite url: the dots are crucial here or it doesn't work – in this case they indicate that the summation goes on forever).

KEEP CUTTING IN 2 PIECES...

ZENO'S CAKE

Alright, we've got cake now, but before we can get this party really started, we need to make some confetti. We will do this as follows: the circle can be seen as a collection of infinitely many points. Identifying each point on this circle with exactly one piece of confetti[28], this means that we can turn our circle into an unlimited supply of paper party pellets. Here's the plan: using a clever argument, we will sort the points on the circle into several heaps – each of which gets a different colour. Well, I say 'several', but that's just a euphemism for 'infinitely many'. Once this is done, we will take some (oh yes, another euphemism) plactic bags, and put one point from each pile in each of these bags. This way, we can turn our original circle into infinitely many bags – each one containing infinitely many coloured points, exactly one from each of the stacks we have created.

---

28   I never thought I'd use the word 'confetto' some day, but here we are!

How do we manage this distribution problem, organising a collection of points on the circle into a bunch of confetti piles? The first point – say the zero angle or the number 12 on the clock's face - marks the beginning of the first pile: it's a dirty job, but someone has to start it. For each of the (infinitely many) other points to be sorted, we *either* have to add it to an existing pile, *or* we have to start a completely new pile for it[29]. 'But how do I know what to do?' Good question, we need some sort of criterion to answer that. This goes as follows: if we pick up a point p on the circle, we can only add it to an existing pile if the difference between p and any number from that pile (it does not matter which one) is a *rational number*[30]. Recall that we are using angles (in radians) to identify points on the circle with numbers, so 'the difference between two points on the circle' is just a number expressing how big the angle between those points is (in the same way like π, or 180 degrees, is the difference between diametrically opposed points on the circle). So it all depends on this angle: if its value is a fraction, we must add p to the pile we have just tested. If not, we move on to the next pile – and so on. Once we have tested all the piles, and none of the differences turn out to be rational (modulo 2π), we have to start a new pile for p and choose a colour for it. Judging from how much time some people seem to spend selecting a colour for their wallpaper, I actually believe that this last part is the hardest thing about the whole procedure.

For instance: all the points on our circle corresponding to the numbers on the clock's face end up on a different pile. The number 3 on the clock (or $\frac{\pi}{2}$ as an angle) and the number 9 on the clock (or $\frac{3\pi}{2}$ as an angle) cannot be in the same pile, as their difference is equal to π and this is not a fraction. On the other hand, the pile containing the angle $\frac{\pi}{2}$ does contain numbers such as $\frac{\pi}{2} + \frac{1}{2}$ and $\frac{\pi}{2} + \frac{42}{37}$. So, as long as we keep adding fractions q (between 0 en 2π) to an arbitrary point from an existing pile (modulo 2π),

---

29   The blue pile or the red pile, Neo, but in a polychromatic version.

30   Modulo 2π of course, as a full rotation does not make a difference. This means that the numbers 3 and $(9 - 2\pi)$, for instance, belong to the same pile.

the criterion tells us that this will give us points on the circle belonging to the same pile. Now, it will take some work – since there are infinitely many points on the circle to be sorted – but in the end we will have an infinite amount of confetti piles (each pile its own unique colour – good luck finding a Caran d'Ache box containing enough crayons).

Okay, are you ready for an example of a non-measurable set? Take a bag[31] and select a unique point from each coloured pile of confetti. 'Are you sure this is even possible?' Well, you can do it if you invoke the axiom of choice. Remember that this whole Banach-Tarski commotion – which has inspired us to consider non-measurable sets in the first place – is one of the reasons why some people seem to refute this axiom, so it was bound to pop up in our argument somewhere. If you are one of those people – choosing not to choose – it will be impossible. But if you are like me, and you accept the idea that you can make a new collection of things (points) by selecting one item from an infinite amount of collections of things (coloured piles of confetti), this poses no problem at all.

However, and this is the upshot of the construction: our bag of confetti will be *non-measurable*. The thing is: it either has a size, or it does not have a size. This is one of the cool things about maths: it's either yes, or no – there are no in betweens[32]. But as we will soon see, assuming that our bag has a size will lead to utter bollocks. To see this, let us have a closer look at the seemingly more logical option that our hand-picked bag of confetti has a well-defined size. Just to make sure, this still means that we have two options left: either the size is equal to zero, or it is a strictly positive number (let us call it S for size). In order to show that both of these options lead to a contradiction, I have to mention two

---

31   Oh yes, we are going to put a set the size of which cannot be measured inside a bag. That's like compiling an Ultimate Top 20 of inaudible sounds.

32   It's like when you punch yourself and it hurts: then you're either too strong, or too weak.

crucial facts. First of all, as the proverb goes: one bag of confetti does not make a circle. Put differently: in order to reconstruct our circle we need more than one bag. I suppose you got used to sentences involving 'infinity' in the meantime, so guess what: you need infinitely many bags of hand-picked confetti to do so. This shouldn't really come as a suprise, since every pile contains an infinite number of points – one for every fraction q that you could add during the sorting process. This is actually how you can make all the other bags: if the first bag of confetti contains for instance the points $\sqrt{2}$ and π, then the other bags must contain $\sqrt{2} + q$ and $\pi + q$ (in a sense, this number q labels the bag). So each bag is 'a copy' of the first bag, up to adding the number written on the bag (modulo 2π).

However, and this is the second crucial piece of information: although this number of bags is infinite, it is the kind of infinity which allows us to apply the additivity rule for measures[33]. Think of Zeno's cake: there will be one bag of confetti for each of the pieces. But this is where the additivity rears its stubborn head: as the circle is the disjoint union of these infinitely many bags of confetti, the size of the circle must be the sum of the sizes of these bags. So there are two possible conclusions here. If the bags have zero size, then the size of the circle must be zero ($0 + 0 + 0 + \ldots$ will still be zero). If, on the other hand, the bags have size S, then the size of the circle must be infinity $S + S + S + \ldots$ gives infinity). However, we knew well in advance that we should get 2π as a result, so this clearly poses a problem.

'So did we make a mistake somewhere?'

Nope, we didn't. What you've just read is a sound[34] mathematical argument, using a formula which somehow seems to work for

---

33   Readers who are not familiar with 'kinds of infinity' will be delighted when they see the next chapter...

34   The Sound of C, to be more precise, the sound which creates a new dimension.

cake. So one possible way out of this conundrum is to give up the additivity of measures, as this played a crucial role in our argument (this basically means that we would no longer be allowed to use the formula). This might come in handy in real life – just imagine cutting up your 21 inch by 16.5 feet roll of wallpaper into 7 pieces, and having enough to decorate 3 African countries – but it would most certainly be the end of geometry as we know it. So instead, mathematicians opted for the only other conclusion possible (I told you, no in betweens), and that was to accept the somewhat strange idea that there exist sets which have no size (like the coloured bags of confetti from above).

2.3.2 Formula

The ultimate representation of most theories is a compact equation, which translates an abstract idea into a string of symbols that nicely fits in a text box. Or a blackboard, for that matter: just think of newspaper articles and movie scenes involving a mathematician — these always come with a badly erased blackboard containing a selection of formulas. Most of the times the displayed formulas are correct, but they often have as much to do with each other as literary analysis rules, verb conjugations and snapshots from an alphabet chart in an English class.

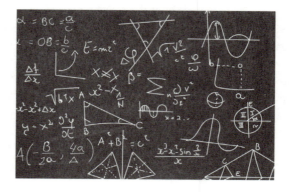

The standard newspaper blackboard

Scientists just *love* formulas, but I have the impression most people don't. Well, I guess they are a bit like incontinent badgers: in principle people have nothing against them, it's just that they don't want them to interfere with their lives. Therefore, it is always dangerous to use them in a book intended for a broad audience. Or as the late Stephen Hawking once formulated it: 'Someone told me that each equation I included in the book would halve the sales.' Or, with S_0 the expected sales and E the total number of equations:

$$S(E) = S_0 2^{-E}$$

And this is probably the most expensive joke in the book.

2.3.3   Owls

A tongue-in-cheek name for the origin in the two-dimensional plane, since it is written as (0, 0) in coordinates — which loosely resembles the face of an owl. One could argue that this shows that some mathematicians do have a non-zero sense of humour. *Nought.*

Genuinely funny science and maths jokes do exist, but for some bizarre reason they often deal with trying to find out why or how poultry got to the other side of the street, with repairing broken artificial lighting systems or with putting inadequately large mammals into fridges.

- How did the chicken cross the road? It read a textbook on topology (a branch of mathematics we will come back to later) and realised it is enough to identify opposite sides.

THE OTHER SIDE

CHICKEN

GLUE THESE SIDES TOGETHER

AND CUT IT OPEN AGAIN

THE TOPOLOGICAL CHICKEN

If this joke somehow makes you think of the quality rating system for movies and TV shows, you may already be familiar with 'the topology of the cylinder' in real-life applications: some programmes are so painfully bad that they become cool again — crossing the spectrum like the chicken above.

- How many mathematicians does it take to change a lightbulb? Two. First of all, the required number M_b is definitely prime: if not, one can divide the mathematicians over d rooms, with d a divisor, and reduce the argument to one room. It then suffices to note that 2 is the smallest prime number.

- How does a mathematician put a rhinoceros in a fridge? He cuts it into five pieces and reassembles these into a pea-sized mammal. He does not even need to remove the giraffe left there by the oblivious physicist from another joke.

Definition 15.

Fridge: *an electric closet in which people tend to collect squeezy bottles and jars containing expired food items, under artificially cold conditions. In order to guarantee success, data samples are selected on a daily basis; bad specimens, the expiry date of which is not due yet, are consumed, whereas good specimens (such as expired Thai curry paste, mustard and pickles) are then meticulously put back in place.*

2.3.4 Theorem

At the end of the day, the practising mathematician's final goal is to prove theorems — timeless truths, cast in stone using her favourite font: the letters of the law. For if maths is the Queen of the Sciences, theorems are the bricks with which her grand palace is built. Deliberately designed to defy both doubt and decay, one on top of the other, constantly creating extra chambers — be they fully functional or only ornamental.

A colleague of the famous Hungarian Paul Erdős (1913-1996) — one of the most prolific mathematicians of the previous century, authoring more than 1,500 articles with more than 500 collaborators — allegedly said that 'mathematicians are machines turning coffee into theorems' (the quote is often attributed to Erdős himself, but he ascribed it to his colleague Rényi). I think I know where he got the idea to compare coffee and theorems: better to avoid them after dinner as they can undermine a good night's sleep, and when they are too concentrated you may find yourself in a seated position for longer than you expected. Because sitting behind your desk and trying to penetrate a poorly written proof is like having a disoriented toddler carrying a cocktail parasol as a city guide: you will find yourself hopelessly lost.

Consider the case of the *abc-conjecture* for example, a problem in number theory which appeals to pure mathematicians as it could pave the way for proving many other famous conjectures. Just like Fermat's last theorem, it is a statement about integer numbers — see also the next box for a more detailed account. Shinichi Mochizuki from Kyoto University in Japan claims that he proved the statement in 2015, but the problem is that nobody really understands his alleged proof.[35] It is a 500-page blurb written in some sort of obscure mathematical dialect that he invented himself — his 'Inter-Universal Teichmüller Theory' has yet to become part of the standard curriculum — which raises the question whether boy still cries wolf when there is no one in the forest to hear his voice.

This philosophical reference is not merely a figure of speech by the way, as it is by no means obvious when a theorem comes into actual 'existence'. Much ink has been spilt about this question, and there are many different schools of thought (Platonism, Mathematicism, Intuitionism and Structuralism, just to give a few crossword-unfriendly examples). I believe it's fair to say that most people adopt the philosophy which says that mathematical entities and truths are eternal and unchanging: they exist in a Platonic world, completely separate from our physical universe. According to this philosophy, proving a theorem is like opening the door to a previously unknown chamber in a vast mental fortress which

---

35   At the time of writing, experts seem to have found serious flaws in his alleged proof.

has always been there. There may exist a formula for π we have not discovered yet, but in the Platonic world it is already a fact. An abstract eternal reality whose shadow is waiting to be observed on the walls of our cave. This is completely different from, for instance, the intuitionists' point of view. According to this philosophy, you can still see maths as a fortress, but — not unlike Kim Kardashian's bubble butt — it is completely man-made, and still under construction. This means that mathematical facts are a creation of the human mind, and just like fairy tales they would not 'exist' if there were no humans to think about them. So a formula for π that we have not yet found, simply does not exist yet. This view on maths does have its advantages for researchers who are afraid that they will run out of ideas in the near future: based on my experience with building contractors, in whose dictionary 'rescheduling' comes before 'punctuality', we might be in for the long haul when it comes to building a fortress.

Note that 'a theorem' should not be confused with 'a law'. The latter is the natural scientist's version of a theorem: it cannot be proved as a thing in itself, but it does express a universal truth. A good example is the law which says that you should never plan an activity with a romantic partner which lies further in the future than the length of your relationship at the time of planning. Remember that time you fell head over heels in love, and you decided to book a summer holiday trip to Italy after merely three weeks of dating? And then you broke up one month later and you had to decide what to do with those tickets? I am sorry for your loss, but you had it coming: that's what you get when you violate a basic law of romance.

The abc-conjecture: in order to explain what this conjecture says, we have to turn to prime numbers once again. Their defining property is the following: all the integer numbers have a unique fingerprint, in that they can be written as a product of prime numbers, raised to a certain power. For instance: $76 = 2^2 \times 19$ and $500 = 2^2 \times 5^3$. Just like your average terms and conditions: exponents are in small print, so why not just ignore them? This obviously changes the number, but it is somehow related to the number we started from. We call the result *the radical* of an integer number, denoted as rad(n). So rad(76) = 2 × 19 = 38 and rad(500) = 2 × 5 = 10. Most of the times, the radical is smaller than the number that you started from — I guess radicalisation rarely produces greater things — but this isn't always the case: rad(30) = 2 × 3 × 5 = 30. So there is an obvious problem here: can we somehow predict *when* it will be smaller, and precisely *how much* smaller it will be. The abc-conjecture does just that: it says that whenever *a* + *b* = *c*, where *a*, *b* and *c* have the special property that they do not have a prime factor in common — just in case this rings a bell: it means that their greatest common divisor is equal to one — then something special can be said about *c*. As a matter of fact, under these circumstances, one has that *c* < rad(*abc*). Well, *most of the times*. To mathematicians, these last words are like staplers at a birthday party: they just don't cut the cake. So is there any way in which we can adjust the statement so that it becomes true? Yes, and of course it all has to do with the exponent. Terms and conditions, they are there for a reason I suppose. As soon as you pick an exponent larger than 1, and it can be as little as $1 + \epsilon$ for a ridiculously small number $\epsilon > 0$, it will hold for all *a*, *b* and *c* provided '*c* is sufficiently large'. I know; when it comes to being vague this description surely competes with 'maths is all about numbers and stuff', but just think of the way some people think about penis size: there is no need to fix a number, it just needs to be sufficiently large.

2.3.5 Isomorphism

Henri Poincaré, the French mathematician (1854-1912) whom we will also meet in a later chapter (when I talk about Grigori Perelman), once said: 'mathematics is the art of giving the same name to different things.' I feel slightly uncomfortable arguing with this polymath, sometimes referred to as 'The Last Universalist' (he excelled in all fields of the discipline as it existed during his lifetime), but I actually believe it is the other way round: 'mathematics is the art of giving different names to the same thing.'[36] And isomorphisms express just that idea: two mathematical objects are said to be isomorphic if, on a deeper level, they are essentially the same thing. So in the land of abstract blindness, isomorphism is the king.

For example: some people enjoy working with cubic polynomials, which are functions of the form $P(x) = ax^3 + bx^2 + cx + d$, whereas others wake up in the morning to get all excited about matrices

$$M = \begin{bmatrix} a & b \\ c & d \end{bmatrix} \ .$$

The former is a polynomial (abstract object number one), the latter is some sort of array (abstract object number two), and yet: they belong to sets which are *isomorphic* as vector spaces (the meta-abstraction). Like 'Lewis Carroll' and 'a gifted mathematician noted for his facility at word play and logic'. Alike through the (right) looking glass.

---

36   Yes, you may quote me on that.

2.3.6   Proof

Regardless of the philosophy, before a theorem is *really* said to exist, experts must establish the correctness of its proof. This is typically done by referees: researchers working in related areas who get the submitted paper from the editor of the mathematical journal to which the author has sent it, and who then review it anonymously (in the sense that the author is not supposed to know who refereed his or her work). Only when they approve of the author's proof, the results may be officially published. So technically speaking, as long as the proof lies on the referee's desk, a theorem's existence is like a Facebook friendship request: pending approval. Over the years, I have learnt that there are other analogies between mathematical proofs and friendship requests: the more famous the sender, the more likely it is to get accepted, and when it comes from a Russian porn star, it is better to proceed with caution.[37]

In order to prove a theorem, there are several techniques. You may have heard of a proof by contradiction (also known as 'reductio ad absurdum'), a proof by lack of space (see earlier this chapter) and possibly even of a proof by induction, but there are many other ways of proving a statement:

- *Proof by abduction:* tell people that the assertion was planted into your brain by alien visitors from a super advanced society in which toddlers are being taught quantum field theory in kindergarten. Although he described it somewhat differently himself, this is more or less the story behind the work of Srinivasa Ramanujan (1887-1920), the Indian autodidact who is known for his vast collection of outlandish mathematical identities and equalities.[38] He said that he often dreamed of the Hindu goddess Namakkal, who presented him with math-

---

37   On the other hand, they do seem to know a thing or two about the satisfying conclusions you can arrive at when working with perfect shapes and bodies.

38   You may recognise him fro the movie *The Man Who Knew Infinity*.

ematical formulas which he then tried to verify upon waking. Ramanujan formulas are as the mathematical equivalent of *surströmming*, the Swedish traditional fermented herring: no matter how repulsive it may look, there is a select group of connoisseurs who can appreciate it. I mean, imagine you jerk awake in the middle of the night, with the formula below projected on your retina: some people consider drugs in less pressing circumstances.

$$\cfrac{1}{1+\cfrac{e^{-2\pi}}{1+\cfrac{e^{-4\pi}}{1+\cfrac{e^{-6\pi}}{1+\ldots}}}} = \left(\sqrt{\frac{5+\sqrt{5}}{2}} - \frac{\sqrt{5}-1}{2} \right) e^{\frac{2\pi}{5}} .$$

Note that the dots indicate that the fraction continues forever, hereby following a simple pattern: one should expect $e^{-8\pi}$, $e^{-10\pi}$, $e^{-12\pi}$ and so on. Formulas like the one above are like the opposite of rhetorical questions: they seem to formulate an answer to questions no one bothered to ask. Note that the letter e appearing here stands for Euler's number:

$e \approx 2.71828182845904523536028747135266249775\ldots$

Just like π, the golden ration and the fear of low flying drunk pianos, the number e is one of these things that seems to be built into the very fabric of nature itself. It keeps popping up, in a variety of problems, both in mathematics and other natural sciences. We will also meet Euler's number again later, when we talk about portable toilets and the quest for the most beautiful formula in science (see chapter 5).

- *Proof by dissection:* buy a goat, and feed it pulverised sheets of paper on which your claim is written. Slaughter the animal using nothing but a compass, a straightedge and a pencil, and organise a ritual offering to the Pythagoreans. I am not claiming that the followers of Pythagoras actually did this, but Pythagoreanism was indeed a cult. I often have the impression that some people believe that mathematicians are still part of a secret society. One which prescribes that we must all recite three theorems plus proof before we go to sleep, that we must travel to Euler's grave at least once during our lifetime and that we must slit the throat of a goat with a protractor when the moon is full on Pi Day, and use its blood to get our favourite number tattooed on our back. For the sake of clarity, this is not the most accurate description of a mathematician: it can also be the blood of a puppy.[39]

- *Proof by example:* according to this highly efficient method, a statement is true as soon as it holds for three cases, which have to be checked explicitly. Not only is this one of the fastest methods around, hence very popular with students, it also leads to elegant proofs for theorems which are much harder to prove using more conventional methods (for instance, 'all positive integers are less than 4').

- *Proof by jurisdiction:* sue everyone who does not believe you. Back in 1897, this almost became reality: the 'Indiana Pi Bill' is undoubtedly the most notorious attempt to establish mathematical truth by legislative fiat. It all started in 1894, when Edward Goodwin — a physician and amateur mathematician — believed that he had found a way to square the circle. This problem, which dates back to the Ancient Greeks, asks for a method to construct a square which has the same area as a given circle, using nothing but a compass, a straightedge and a pencil. It was suspected to be impossible since ancient times and rigorously proven so in 1882, but Goodwin clearly was a non-believer. He even proposed a bill to the state rep-

---

39   Some students are being thaught that every time they divide by zero, a puppy must die. This is why.

resentative,[40] in which it was specified that royalties had to be paid to him whenever his 'new and correct value for π' was used. Technically speaking, the name of the bill is a bit of a misnomer — it does not purport to settle a value for the number π — but Goodwin's alleged solution for the squared circle did imply an incorrect value (it leads to π being equal to 3.2 or 4, depending on how you interpret his arguments). After much public ridicule, it was tabled by the Senate. The Guinness Book of World Records lists this legislative lunacy as 'the most inaccurate version of pi'. Recently, a Twitter user going under the moniker of Sir Michael must have thought that this was an easy record to break. He sent a hilarious proposal to the organisation and thereby earned himself the following enviable rejection letter.

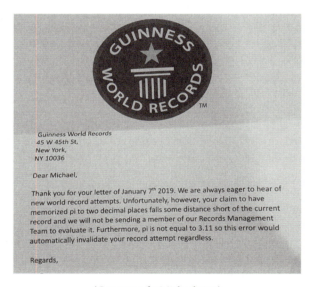

(Courtesy of @Michael1979)

---

40   'A Bill for an act introducing a new mathematical truth and offered as a contribution to education to be used only by the State of Indiana free of cost by paying any royalties whatever on the same, provided it is accepted and adopted by the official action of the Legislature of 1897.' (from 'History and oddities of the number pi' by Vincent Mallette)

- *Proof by obscure referencing*: tell people that you read about it in an ancient paper, on pages made out of human skin, written in golden ink, using a Polynesian language which is now only spoken by 37 remaining people and a parrot, and that it appeared in a journal which is no longer in print because the publishing company got struck by a meteorite.

- *Computer-assisted proof*: as the name suggests, this is a proof in which a computer program was involved. It is sometimes used when the proof can be reduced to performing lengthy calculations or to verifying all possible combinations of a discouragingly long list, or just because it's a sunny Friday afternoon, and the authors feel like 'Screw this shit, we've worked enough this week, let's get hammered.' A famous example of a computer-assisted proof is the one that established *God's number*, back in 2010. This number is defined as the maximum number of moves required to solve a Rubik's Cube, given any of the 43,252,003,274,489,856,000 possible combinations one can start from. It turned out that God's number is equal to 20. Atheists consider this to be yet another proof for the fact that God does not exist: it should have been 42 of course.

The answer to life the universe and everything =

$$42$$

Rad	▦	x!	(	)	%	AC
Inv	sin	ln	7	8	9	÷
π	cos	log	4	5	6	×
e	tan	√	1	2	3	−
Ans	EXP	xʸ	0	.	=	+

Even Google says so.

2.3.7   Conjecture

Before a theorem becomes a proper theorem, it is a conjecture. And it stays that way until someone either finds a sound proof, or a counterexample — the scientific equivalent of a party pooper. From this point of view, Fermat's famous last theorem was not even a theorem until 1995.

One of the oldest, best-known and simplest conjectures which still stands today is the Goldbach conjecture, which dates back to 1742 (or 1723 + 19). It says that *every even number bigger than 2 can be written as the sum of two prime numbers.* Despite considerable effort, including an explicit verification up to integers consisting of 18 digits and cracking as many fortune cookies — the truth may be in there — it remains unproven until today. As such it genuinely gained its place in the Guinness Book of World Records, as the longest-standing maths problem. I am assuming here that the Mysterious Messiah Multiplication — 5 loaves times 2 fish equals 5,000 people — does not count as an algebra problem then.

If only I had known the difference between theorems and conjectures when I was a kid. I remember thinking that 'pictures are worth a thousand words' was a well-established truth. It wasn't until that day I had to write a 2,000 word essay about my summer vacation that I realised that it was nothing but a loose conjecture. After handing in two holiday pictures, the resulting zero score turned out to be my first home-made counterexample. As I never really got over that experience, I have been looking for a proof ever since. A few years ago it finally dawned on me: some pictures *are* worth a thousand words, as you can verify yourself at the end of this chapter.

2.3.8 Trivial

Mathematicians often claim that something is 'trivial' (or obvious). What is meant by that depends on the person using it. My personal experience with that phrase is that it can mean anything from 'you are a complete simpleton if you don't see this' through 'it took me eight days, a container of Red Bull, two boxes of Kleenex, 17 pizzas, four sessions with a psychiatrist and half a ton of scrap paper to verify this, but it could be that I overlooked the easy way to prove it, so in order to save my face I will say it is trivial' to 'we have absolutely no clue how to tackle this problem, but we consulted a Magic 8-ball to know whether it's true and "without a doubt" was the answer'. There is a famous anecdote about a professor[41] who once said 'Of course, this result is obvious.' A brave student raised his hand and said, 'Excuse me, sir, but I don't see it. Is it really that obvious?' The professor stared at the board, and 42 minutes later he said, 'Yes, it is obvious.' In any case, it is trivial to see that you have to proceed with caution when you bump into this word.

2.3.9 Corollary

Most theorems are followed by corollaries. These often are — and I did warn you about this — 'trivial consequences' of the preceding theorem, and can thus be seen as mathematical versions of sentences starting with 'Oh, and by the way'. Not only theorems, but also laws can have their corollaries. Just think of Murphy's Law, the famous epigram which is usually stated as 'anything that can go wrong, will go wrong'. Typical corollaries of this law include 'you never find something until you have replaced it', and 'you always get the most of what you need the least'. Even in maths, there are quite a few notable corollaries:

---

41   As with most anecdotes, each version has its own subject: many people seem to have had 'this professor'.

- The solution to a problem always comes with a minus sign when you don't need one.

- Each time you redo a calculation, you find a slightly different answer.

- The more pleased you are after proving a statement, the more elegant and shorter your colleague's argument will be.

- The more important that one seminal paper is in order to understand a crucial part of the topic you are working on, the more likely it is that it appeared in an obscure journal which is out of print.

- The longer you are working on a calculation, the more likely it becomes that your final conclusion will be that $0 = 0$.

- Changing the order of calculating an integral and taking limits is always allowed when you painstakingly make the effort to check the conditions explicitly. That one time you decide to skip the details, though, it will produce a mistake.

Murphy's Law should not be confused with Murphi's Law — a formula that does not contain the golden ratio ϕ, Euler's number e or number π is most likely wrong — or with the 'Law of Conservation of Misery', which says that the total amount of misery is always a constant. So whenever you try to decrease the amount of misery in one aspect of a system, you will increase it in at least one other aspect. Mathematicians are probably the only kind of people who feel rather comfortable about this law: according to my former PhD supervisor, the whole idea behind good research is that for each problem that you can solve, at least two new problems should be created. This may sound like 'one step forward, two steps back' to many people, but I know at least one other guy who turned this philosophy into a goldmine — or am I really the only one who reads this as the instructions to do the moonwalk?

2.3.10 Elegance

One of the most peculiar features of mathematicians is their ability to judge the 'elegance' of abstract reasonings. Unfortunately — not unlike *The Matrix* — no one can be told what it is: one has to see it for oneself, as there is no clear-cut description. The mathematical elegance of a proof sometimes boils down to it being surprisingly succinct, or the fact that it connects existing results from previously unrelated fields.

Whatever it is, it seems to come with experience: I noticed that PhD students — and that includes me from 15 years ago — tend to stick to hardcore calculations when they are trying to prove something. However, it often turns out that an 'easy argument' — one which requires you to look at the problem from at least three different angles before you start digging in, not unlike some people's attitude towards seven-course meal plates in Michelin-starred restaurants — allows you to reach the same conclusion, condensing 16 pages of computations into a single paragraph. When it comes to proving something, calculations should be the mathematician's last resort — like flying for street pigeons: I have seen more than one of these birds stubbornly trying to cross the road on foot, as if they're being penalised every time they use their wings.

Paul Erdős, whom we met earlier, even had a name for this. Although he was an avid atheist, he did believe in 'The Book': a dog-eared, coffee-stained volume with a train ticket bookmark — let us not forget that we were created in His image and likeness — in which God supposedly keeps the most elegant proofs for theorems. Whenever he saw a particularly beautiful proof for a mathematical theorem, he would exclaim, 'That one is from The Book!'[42] I find this very odd, an avid atheist who believes in some kind of Holy Script, but what some people find even stranger is that he constantly travelled around the world, visiting colleagues

---

42   See Proofs From The Book by Martin Aigner and Günter M. Ziegler.

and living out of his suitcase. To be honest, I don't know what is so remarkable about that: inside a suitcase it's too dark to do maths anyway.

Definition 16.

Book: *an archaic prototype of traditional reading devices such as tablets, mobile phones and e-readers, containing printed versions of webpages. Unlike modern tablets, these are to be swiped horizontally — from right to left. They can be physically dowloaded from giant servers called 'libraries'. Their only advantages over modern-day devices are their battery life (nearly infinite), the quality of their front screen (as good as unbreakable) and the fact that they can be used to swat mosquitoes.*

So it turns out that mathematicians not only want to solve problems, they even want to do it as elegantly as possible. This makes me think of cartoon figures, and their way of dealing with a falling tree: not only were they *always* in the plane spanned by the standing tree and its final resting position, which is an improbable miracle in itself, they also could never resist the temptation to start running away in the same direction as the tree. Am I really the only one who thought that just stepping aside is a more elegant solution?

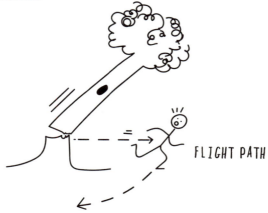

FLIGHT PATH

SURELY THIS MUST BE SHORTER, RIGHT?

2.4 A bleeping theorem

Earlier in this chapter I mentioned a proof that I recently discovered, one that alleviated the pain inflicted on me by a zero score for my school essay, back when I was already old enough to watch cartoons with a critical eye. Well, I guess a promise is a promise.

Theorem 1. *Some pictures are worth one thousand words after all.*

Proof: Before I moved to Antwerp — the city where I currently teach at the university — I used to be a train commuter.[43] I have always enjoyed train rides. Being on the move seems to put me in a state of mind which encourages me to do research. *Mens rapida in corpore rapido.*

One day I was scribbling down a few ideas in my notebook — hiding behind a giant set of headphones, the socially accepted equivalent of hanging a sign with '*Do Not Disturb*' around your neck — when I noticed a rather eccentric woman making her way down the aisles. Breaths were held, empty seats suddenly claimed by bags. I did not have a joker card to play, so I ended up getting the company of this older lady, wearing a park ranger's type of hat and matching boots, beaming at me through an old-fashioned pair of glasses resting on the tip of her nose.

<bleep>

My nutty people detection device was triggered. 'Excuse me, what are you doing?', she suddenly asked me in an English accent that sounded as if it was still recovering from a recent fracture.

---

43 In mathematics, an operation is said to be *commutative* if the order doesn't matter. Addition, for instance, is commutative as $A + B$ is equal to $B + A$. Given the fact that going from A to B is essentially different from going from B to A, I have always found this a weird choice of words in the context of travelling. I surely anti-commute to work.

I decided to ignore her. I did not really feel like engaging in conversation with a retired fairy from a distant forest lodge.

'Excuse me,' she repeated, 'what are you doing? Are you writing a poem?' I lifted my headphones and decided to play the business card. 'I am working, madam. So if you excuse me, I would rather continue writing.'

'What are you working on?', she wilfully continued, clearly mistaking my polite retort for an invitation to start a conversation. 'Mathematics,' I tried, desperately hoping that this would help to make her understand the gravity of my work.

'Interesting,' she replied, extending both arms. 'Can I have a look?' I gave her my calculations. If luck was on my side, she would lose herself in them. She stared at my handwriting like someone finishing a 10,000 piece jigsaw version of the Taj Mahal: completely puzzled out.

'This does not look like what I have have learnt at school,' she concluded. 'Of course not,' I thought, 'this did not even exist back when you were at school. That is why they call it current research.' I remained silent though, aware of the people on the other side of the aisle staring at me as if I were the oddball.

<bleep>

'What have you all become,' she muttered, 'plenty of people on this train, and yet nobody feels like talking to me.' I felt cornered into my discomfort zone. 'I say you have all become socially handicapped.'

I could not help but look up and was met by a mischievous smile. I did not want her to have a point, so I decided to unplug my iPod and surrendered.

'Why do you need music for inspiration?', she asked me — picking up the threads of a conversation we weren't having in the first place. 'Why not just look outside?'

'It does not work that way,' I said. 'It is not like I can draw inspiration from that.'

'I bet you can,' she nodded. 'Look, I will show you what I am carrying to Antwerp, maybe it can trigger you to write something.' From her faded hippie bag she took a bunch of flowers, a plastic bag containing potatoes and a brown paper bag full of chestnuts. If she wasn't, she at least brought them along. Nuts.

<bleep>

'Flowers, cooked potatoes and steamed chestnuts.' She somehow succeeded in making this sound self-evident.

'And what are you going to do with that?' The stares from the other side of the aisle did not bother me any longer.

'I am visiting friends whom I haven't seen for 25 years. This is their present.'

'Where do they live?', I asked, trying to imagine how I'd feel if someone were to give me these presents. After 25 years.

'No idea, I will find them when I get there.'

<bleep>

She took three pictures from a crumpled envelope. 'My biggest surprise for tonight,' she said, handing them to me. 'I found these at home, in a box in the attic,' she added. One was a group picture, which was so blurry and out of focus that I almost expected to see the Loch Ness monster in the background. 'This was when the television crew visited our village,' she explained. It turned out that a film crew had once shot a scene in the place where she was born. 'This man,' pointing at a murky face in the background, 'is my best friend. It is him I am going to see.'

'And those people?', I wanted to know, showing her the other pictures. 'Those are his parents.'

A short silence nested itself comfortably between us. My encounter with this odd-looking lady — eagerly elucidating her endeavour to pay people she had not seen in 25 years a surprise visit — had become a crash course in nuclear physics: random collisions can create energy.

'Do you mind me taking a picture of you?' she suddenly asked me. Had she not just shown me these pictures of the friend she was visiting, I would have said no. Call it vanity, but the prospect of a faded picture with my face on it sparking a conversation on the train one day made me say yes.

'Funny,' I blurted out, 'you should come to Antwerp again in 25 years.' She looked at me sideways. 'Yes,' I continued, 'when you find this picture you just took in a box in the attic. Then you can take the train to come looking for me.'

She stared at me in an almost pitiful way. 'You do not have to look for things,' she declared. 'You just find them.' Prophetic words, because professional experience has taught me that some of the best ideas I have had came when I least expected them.

She was right: the potatoes, the chestnuts and the flowers did indeed inspire me. For a thousand words.   Q.E.D.

I don't know whether you even bothered verifying whether the theorem above really contains exactly 1,000 words, but you can count on my word(s): being a mathematician, I can do basic arithmetic if I really have to — not as good as an adjudicator though — and that includes counting up to 1,000. This being said though, I would like to use this opportunity to debunk a popular myth: mathematicians do not particularly *enjoy* doing arithmetics.

It's somewhat like with proctologists and haemorrhoid removal: yes, they are probably better at it than you are, but that doesn't mean they actually enjoy doing it.

So next time you find yourself in desperate need of someone to do the analytical continuation of that Dirichlet series you dug up from between the cushions of your sofa, do not hesitate to call on us. But can you please stop expecting us to do the division when we decide to split the bill? It's not like we lose our ability to divide by 7 after a night of downing beers with six friends — there is a very easy trick to do this in your head: calculate 13 times the given number, then multiply the result by itself, divide it by 8,281 and take the square root — we are just sick of being stereotyped that way. If you really want to stick to clichés, why not give the bill to your friend the philosopher, who will happily engage in a long-winded conversation with the waiter to convince him or her of the fact that it does not even exist from a metaphysical point of view? Or when you are having dinner on a Friday evening, you can always pass the bill to your friend who works as a public servant: he can then calmly explain that he cannot do anything about the situation because it happens to be after 3pm already, and the financial transaction will therefore have to wait until next Tuesday ('I am afraid Monday is a Bank Holiday'), provided the required forms are filled in and brought along, together with three recent photographs, a bank statement translated into four languages, a certificate of good conduct, seven copies of the medical records printed in dark red capitals on canary yellow A5 pages and an autograph from two former history teachers.

Recommended listening

Artist	Song title
Endzweck	Proof
Hatebreed	Proven
The Black Heart Rebellion	Leaving the Capitals
Korn	Word Up!
Eminem	Difficult (tribute to Proof)
The Dillinger Escape Plan	Chinese
Gloria Jones	Tainted Love
As Friends Rust	Coffee Black
Dropkick Murphys	Out of Our heads
The Cure	Pictures of You

3

Setting Up Space

Either mathematics is too big for the human mind,
or the human mind is more than a machine.

(Kurt Gödel)

You should really read this chapter if ...

- you want to know why there is no such thing as one special set that rules them all.

- you don't want to book a night at the Hilbert Hotel.

- you are not really sure whether you should shave or get shaved.

- pirates intimidate you.

3.1 Maths you can count on

So let us get back to that question we mentioned in the first chapter: how do mathematicians define 'a space'? As mentioned there, spaces are sets to which some sort of structure is added. Before we can turn to structure, let us therefore have a look at these things called 'sets'. They are a bit like molecules for professional cooks: they are constantly being used on a fundamental level, but hardly anyone ever realises this anymore, or simply cannot be bothered. Some mathematicians obviously do care about them, a lot, and do not make a secret out of it. We call these people the set theorists; they are the molecular cooks of mathematics: their topics of interest include the very foundations of abstract reasoning, a subject which lies at the crossroads of maths, philosophy and hairsplitting using a pencil.

Definition 17.

Philosopher: *a special kind of human being, whose enquiring nature is built into the genes: in sharp contrast to males and females, philosophers have nothing but Y chromosomes. They are genetically predisposed to suffer from an existential crisis (Why am I?), sometimes combined with a multiple personality disorder (Why are we?) or an uncontrollable urge to incorporate hip-hop slang (Y is I?).*

Nowadays, set theory is no longer part of the standard maths curriculum, but when I was a kid it was actually taught at primary school: I remember it being a tough subject. Not only because we were expected to understand what formulas like

$$A \cap (B \cup C) = (A \cap B) \cup (A \cap C)$$

meant, but also because our teacher used to deduct points for poorly drawn Venn diagrams. That is like giving fewer points to synchronised divers because you do not like their swimsuits.

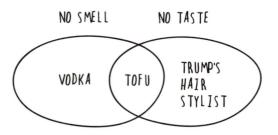

Example of a Venn diagram

At this point you may be asking yourself, 'What exactly is this thing you call a set?' Well, mathematically speaking a set is nothing but 'a collection of objects which belong together', often visualised as a bunch of items contained within a Venn diagram. Just to make sure that you get the idea behind this definition: a set *may* contain sandals and white socks, although we all know these do not belong together. Well, most of us do.

Definition 18.

Sandals: *a rather special kind of footwear, often worn by mathe-maticians and people who believe that this will help to reduce their ecological footprint.*

At first glance, sets seem to be a bit dull. You can give them a name and you can verify whether a certain object belongs to a set or not — expressed in terms of that handicapped euro sign '∈'— but that is about it (so given a set A and an element a, we either have $a \in A$ or $a \notin A$). Until you pick up that shovel and start digging, because sets actually lie at the core of some of the most memorable revolutions in mathematics, and we will come to this in a moment.

Let us start from the obvious observation that there is at least one other thing that you can do with sets: you can count how many elements they contain. It can be zero (the number of vowels in the Czech sentence '*strč prst skrz krk*', which means 'stick your finger through your throat' — probably the best way to pronounce it), it can be 25 (the number of people that crammed themselves into a phone booth in Durban, South Africa, on March 20th 1959[44]), or simply too much (the number of raisins in a bag of trail mix, or am I really the only one who thinks the nuts are always outnumbered?).

In a sense, as long as the number of elements is *finite*, it basically amounts to kindergarten skills: counting. Which does not mean that it cannot lead to interesting mathematical questions. Numbers are like holes in the ozone layer: once they start getting too large, certain issues arise. In a classic joke, two people challenge each other to name the bigger number. The first person, after a few hours, triumphantly announces, 'Thirty-seven!' The second ponders for a while, and finally concedes 'You win.'

---

44 According to Wikipedia, no one could not answer the phone when it rang though. One can only hope it wasn't the adjudicator calling, asking them to confirm their Guinness World Record submission.

This may have started off as a joke, but it did become serious business at some point: mathematicians were trying to come up with *seriously* large numbers, figures that make your head spin. So we are not talking about *astronomically huge* numbers here — things like the number of cells in your body or the number of neural connections in your brain (oddly enough both equal to 10^{14}, give or take), the number of atoms in the observable universe or the number of emails I have received over the past few years from Nigerian princes, asking me for my bank account number because I turn out to be a royal heir of some sort (also roughly equal, give or take 10^{80}). All these numbers are still 'manageably small'. Well, let's not exaggerate: make that 'manageably big'. One of the reasons for this is that we can still write them down in a very compact way, using the exponential notation.

This amounts to using powers of 10 to denote large numbers, which is often done in science. These numbers are of the form 10^k, with k a number which then tells you how many zeroes to add after the initial number 1. This is obviously only true when k is a *positive* integer, but because we are using them to describe huge numbers we will ignore negative exponents k. Consider the number *Googol* for instance, a term coined by the nine-year-old nephew of the American mathematician Edward Kasner, who was asked to come up with a name for a very big number. Googol is equal to 10^{100}, or the number 1 followed by 100 zeroes. Larry Page was fascinated by this term, and later used it for the well-known search engine. He thus clearly ignored it when it asked him *'did you mean Googol?'* though, as he misspelled it. This number can still easily be written down explicitly: I just tested this on a piece of scrap paper, and it only took me three minutes and 17 seconds to do so. To be completely honest, I did get bored after half a minute and I decided to update my Facebook status, made myself two sandwiches, loaded a washing machine and replied to 73 emails. I almost reached that supreme level of efficiency I can only attain when brushing my teeth, which amounts to wedging my tooth-brush between cheek and teeth after a few seconds, and finishing tasks that would normally take me half a morning in less than three minutes. Seriously, I think companies should encourage their employees to have dental hygiene breaks instead of cigarette breaks: better for personal health and the economy. The only side remark is that we may have pushed it too far with the electric toothbrush: some say that people are writing fewer letters since the advent of computers and email, but I noticed that I am even sending less emails since I bought an electric toothbrush. It is just too heavy to keep it dangling from my mouth.

Even bigger than Googol is the number *Googolplex*, which is defined as 10 to the power Googol. It is thus equal to the number 1 followed by Googol zeroes. Do yourself a favour, and do not attempt to write down this number explicitly. If you are up for a proper challenge, why not translate a Haruki Murakami novel from Japanese into one of Tolkien's Elvish languages, or try to catch a white shark using nothing but two violins and a piece of raspberry pie? Just try anything but writing down Googol zeroes: even if you were capable of writing down ten zeroes *per second*, it would still take you 10^{99} seconds. If I tell you that the age of our universe, measured in seconds, is equal to 10^{17} (13.8 billion years) you may get a hunch as to why I strongly discourage you from doing so.

'Surely this is a huge number then?', I hear you thinking. Well, not really: Googolplex still classifies as 'astronomically huge', so it is far from being one of those *seriously* large numbers that can make your head spin harder than a neutron star in a washing machine. For one thing, Googolplex can still be written down using seven numerical symbols only, if you are allowed to use exponents. In the formula below, for instance, I can use a so-called exponential tower of length four:

$$\text{Googolplex} = 10^{\text{Googol}} = 10^{10^{100}} = 10^{10^{10^2}}$$

The massive numbers that I am talking about have so many digits
that even a whole army of monkeys on typewriters would not be
able to reproduce their decimal representation (then again, unless
your tactic is to undermine your enemy's administrative support
by wearing out its hardware, such an army is rarely useful). There
exist integer numbers which are *so* immensely big that you need
a whole new machinery simply to *describe* them, because even
towers of exponents fail to capture their size. Numbers which
require things like hyper operators, Ackermann functions, Knuth's
up-arrow notation or Conway's chained arrow notation — techniques which are by no means part of the standard mathematical
jargon. It would lead me too far astray to even attempt to explain
how mindnumbingly huge a number like $5 \uparrow\uparrow\uparrow 19$ actually is, but
the idea behind Knuth's up-arrow notation is a bit like the scoring
philosophy behind professional darts: the more arrows you can
use, the higher your score will be.

The integers that can be constructed using these concepts nicely
illustrate the power of abstract mathematics: not only have we
built the pyramids, sent people to the moon and designed a dress
that had the power to divide the inhabitants of Planet Internet into
two competing camps (those who thought its colour was black
and blue, and those who were convinced it was white and gold[45]),
we also evolved into creatures that can *imagine* elementary entities — as simple as 1 2 3 but then 'a bit' further down that line —
which cannot be written down in explicit decimal form. Not even
if you are allowed to use towers of exponents, because there will
simply not be enough space, nor enough time in the history of the
universe.

And yet, we have conceived these numbers, and been able to say
useful things about them. Well, I say 'useful', but what I really
mean is that these gargantuan numbers sometimes appear in
mathematical proofs for theoretical results. One of these monsters
even got its own name: 'Graham's number' G_{64} appeared in the

---

45   See https://en.wikipedia.org/wiki/The_dress.

Guinness Book of World Records in 1980, as the largest integer number which was ever used in a mathematical argument.[46] At that time, that is, because since then Graham's number G_{64} has been dwarfed by several other giants. Even I was stunned when I found out that G_{64} (it's just a non-transparent name for a number, in the same way as HR 5171 Ab is the official name for a large star in our universe) is effectively peanuts compared to the number TREE(3), for which I happily refer you to the interwebs.[47]

3.2 A busload of infinity

Given the fact that mathematicians have invented this whole range of new symbols with which they can create numbers that you cannot even pronounce properly — let alone calculate — you may feel like you will never be able to win this 'name the bigger number' competition against mathematicians; like watching cricket players when you grew up with baseball, they seem to be playing a completely different ball game. So you might be tempted to start using the word 'infinity' in your answer — surely that road must lead to victory, right? I hate to burst your bubble once more, but you could not have been further from the inconvenient truth. In the words of the American author John Green: 'some infinities are bigger than other infinities'.

This was essentially the surprising conclusion Georg Cantor (1845 -1918) arrived at in the late 19th century: there exist — fittingly enough — infinitely many *different* infinities.

---

46 Not to be confused with the biggest number ever used in a relationship argument, which is undoubtedly a gazillion ('I already told you a gazillion times, but you never seem to listen to me!').

47 Readers who are really interested in ridiculously huge numbers can always search for David Metzler's YouTube series about this topic. A grand (well, considering the size of the numbers he describes: not really) total of 46 episodes, a real recommendation.

In his seminal paper from 1874 and a subsequent series of papers in *Mathematische Annalen* (one of the most prestigious journals), he worked out this hierarchy of infinities and in doing so laid down the very foundations for set theory, the topic of this chapter.

Fathering a new discipline is obviously quite the honour, but as is often the case when a paradigm shift occurs, Cantor was not recognised for his fundamental breakthrough at that time. Some colleagues called him 'a great disease' and 'a corruptor of youth', putting him in the same league as techno music and Pokémon Go. His work was regarded as so shocking that it provoked objections from mathematicians, philosophers and even Christian theologians. They got involved in the censure because they argued that his theory challenged the absolute infinite nature of God. This criticism of his work weighed heavily on him, and Cantor is known to have suffered from several bouts of depression.

Given my own brief encounter with depression, seeing the bitter irony in this episode of mathematical history pains me. For most mathematicians, and I surely form no exception, the attractiveness of our discipline precisely lies in the fact that it embodies absolute, timeless certainties. For me, this timelessness even provides a safe haven in periods of emotional turmoil: in the maths department, there is no need to think about solutions before the problem arises, nor to make a problem out of resolved issues. I am in no way claiming that spending too much time in the mathematical realm is good for you — not least because it may affect relationships, as I have found out over the years — but merely knowing that there is at least one place where our thoughts can be in perfect control of the situation means a lot to me. If I then think back to Cantor, who did not even find rest there in that foolproof world, with its eternal truths — it makes me realise that I do not aspire to achieve a breakthrough if it comes with a breakdown.

On a more positive note: Cantor did receive accolades for his work before he passed away. In the words of the great David Hilbert: 'From his paradise that Cantor with us unfolded, we hold our breath in awe; knowing, we shall not be expelled.' He is now widely regarded as a sheer genius. I suppose it is only fitting that he who unleashed the infinite will be associated with it forever and ever.

So what exactly is this hierarchy of infinities then? The 'simplest' infinity is related to a set we already encountered in the previous chapter, namely the one containing the natural numbers:

$$\mathbb{N} = \{0, 1, 2, 3, 4, 5, \dots\}$$

Note that some people adopt a different convention, in which 0 is *excluded* from the set of natural numbers (the positive integers). I suppose these people also tend to confuse first floors with ground floors.

The set $\mathbb{N}$ contains infinitely many elements, and the 'size' of this set is denoted by $\aleph_0$ (read: aleph-zero). It should be emphasised here that this is not a number, it is simply a symbol representing a certain amount — which happens to be *a lot*. The number 7 is obviously a number, but in a sense it is also simply a symbol representing a certain amount: all sets containing seven elements have this symbol in common. The one containing the deadly sins, the days of the week, the pure notes in the diatonic scale, the liberal arts, the colours of the rainbow, the number of times I have read *The $13^{1/2}$ Lives of Captain Bluebear:* one could almost say that 'seven-ness' is a property that they share. One can check this by counting the number of elements, but surprisingly enough this is not the only way.

Even without explicitly counting the elements in these sets, we can still easily see that they all have the same amount of elements, by simply pairing them up, two by two. This seems trivial, but that's because you're too good at counting. From a conceptual point of view, pairing them up is different from counting. Think of envious rich people in an upscale part of town: they are never really interested in *how many* cars their neighbours have, they only want to have more.

And to accomplish this, there is no need to count: as long as their line of cars is longer when they put them next to each other, they have more cars.

2 SETS WITH THE
'SEVEN-NESS PROPERTY'

A bijection between deadly sins and days of the week

Mathematicians call this act of pairing up the elements in such a way that nothing gets left behind '*defining a bijection between two sets*', or — when they are in a less verbose mood — 'connecting dots'. Back when your kindergarten teacher taught you to count those five apples on the table, she essentially asked you to construct a bijection between the set of apples on the table and the set of fingers on one hand. She most certainly did not phrase it that way — because she knew that you had other things on your mind, like planting pieces of snot under your chair — but that's what counting boils down to: setting up a one-to-one correspondence between two sets. For finite sets this seems like overcomplicating stuff, but it turns out to be utterly crucial when the sets contain an infinite number of elements. However, it leads to the type of surprising conclusions that made Cantor's work so contested.

To see why the case of infinite sets is different, let us consider an example. Suppose there are $\aleph_0$ buses, which thus means that the number of buses is in one-to-one correspondence with the natural numbers. You can't count this number — it's infinitely many — but you do know that for each integer $k \in \mathbb{N}$ there is a unique bus. You can literally line them up, on an infinitely big car park. Next, suppose each bus contains $\aleph_0$ passengers: one bus driver, a handful of teachers and enough teenage school kids to prove that you don't need to see Jeremy Clarkson fulminate about the Lexus SC 430 (the winner of *Top Gear*'s Worst Car in the History of the World) to know what 'a hell on wheels' means. Once again, you can't count the number of people on a bus (there are infinitely many), but you do know that the number of people is in one-to-one correspondence with the natural numbers.

Let us then start labelling things: every bus has a unique natural number $k \in \mathbb{N}$ on its roof, and every seat on each of these buses has a unique number $\ell \in \mathbb{N}$. Put differently, every person is labelled by two numbers (k, ℓ), the first of which refers to its bus and the second to its seat on that bus. For instance: $(5, 0)$ stands for the bus driver on the sixth bus (don't forget that we start counting from zero). Now here is the question: *which set is 'bigger', the set **B** containing all the buses, or the set **P** containing all the passengers?* The former has $\aleph_0$ elements, one bus for every number, but the latter surely seems to contain way more elements, right? We have $\aleph_0$ buses, and $\aleph_0$ people on a single bus: *surely* there must be 'more' people than buses. We are dealing with infinitely many times infinitely many people, or in mathematical symbols: the set **P** with all the passengers has size $\aleph_0 \times \aleph_0$. And yet, both sets are of size $\aleph_0$. *They contain 'the same number' of elements.* There are 'as many' buses as people.

'Surely you must be joking now, right?'

I know, this seems wrong because we feel an urge to *count* the elements. But this is impossible as both sets are infinitely big.

The assertion is correct though, since there is an easy way to define a bijection between both sets. Just look at the picture below:

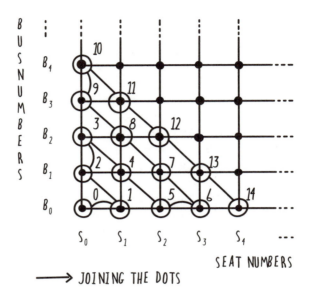

A bijection between **B** and **P**

I actually remember using this technique in kindergarten: hammering a few nails into a piece of wood, connecting these nails using a piece of string — weaving it into some kind of dreamcatcher-like pattern — and then giving it to my dad on Father's Day. It's essentially the same trick: it requires an infinite piece of wood and a longer piece of string, but the upshot is that you *only* need $\aleph_0$ nails. We started labelling our passengers with two natural numbers, but what the picture shows is that this is absolutely not necessary: each person can uniquely be identified using a single number only (its position on the string of nails, so to speak).

Definition 19.

Father's Day: *although the precise date may vary from country to country, this is the day on which fathers and fatherhood are celebrated all over the world. It is but an example from a long list of days in a year which are devoted to celebrating a person, a historical event, an idea and so on. Circled in bright red on some calendars is March 14th: Pi Day for those who are irrationally in love with maths, but it also happens to be Albert Einstein's birthday. Every once in a while, new days are also added to the calendar; some people are trying to install the National Indecisiveness Day, which will probably be held the second Friday, third Wednesday or last Tuesday of May.*

The concept of $\aleph_0$ and its counterintuitive properties are often illustrated in terms of the Hilbert Hotel, named after the famous German mathematician David Hilbert (1862-1943) mentioned earlier. This hotel, with its $\aleph_0$ doors, can easily accommodate each of the $\aleph_0$ people on the $\aleph_0$ buses: it suffices to send the person on seat ℓ in bus k to the room with number $n = 2^k 3^\ell$. The bus drivers, for instance, go to rooms 2, 4, 8, 16 and so on: their rooms are of the form $n = 2^k 3^\ell$ with $\ell = 0$, and you may recall that $3^0 = 1$ so that only $n = 2^k$ remains. Since natural numbers all have a unique prime factorisation, everyone will end up in a different room — until the teachers finally go to sleep of course, because after that the boys all turn up at the girls' rooms to spin bottles and twist tongues. Surprisingly enough, it gets even weirder than that: we can accommodate $\aleph_0 \times \aleph_0$ people in the hotel, with its $\aleph_0$ rooms, but this will leave almost all of the $\aleph_0$ rooms in the hotel empty: we have used the numbers 2 and 3 in our room assignment algorithm, but we have not been using any of the infinitely many other prime numbers (room numbers with a factor 5, 7, 11 and so on), which means that all these rooms are still available. Even for your favourite natural number k (extra points are gained for $k = 42$), one still has that the k-th power of infinity gives infinity:

$$\underbrace{\aleph_0 \times \aleph_0 \times \aleph_0 \times \aleph_0 \times \ldots}_{k \text{ times}} = \aleph_0^k = \aleph_0 \ .$$

However, multiplying with infinity is a bit like going out on a first date: if you become too enthousiastic, things may start going awry. The thing is that if you keep multiplying with $\aleph_0$, on and on and on, you *will* get a new infinity. In terms of our fancy symbol for infinity, we thus have that

$$\underbrace{\aleph_0 \times \aleph_0 \times \aleph_0 \times \aleph_0 \times \ldots}_{\text{infinitely many times}} = \aleph_0^{\aleph_0} \neq \aleph_0 \ .$$

If this formula bothers you, you can take comfort in the fact that you are not alone. I mentioned before that Cantor's colleagues were shocked by his brutal claims — well, this one got all the controversy started. In a sense, all we have been doing so far is exploiting the power of bijections to prove that some infinities are *equal* to others, although they may look 'bigger' (think of the buses and the passengers). Cantor was the first to prove that this is not always the case: some sets can indeed contain 'more' elements than $\aleph_0$, the 'amount' of natural numbers, making them 'bigger infinities'. Using his famous *diagonal argument* which was later named after him (see the box below), he discovered the existence of a 'bigger infinity' $\aleph_1$ which is now referred to as 'uncountably many'. This is a bit of a misnomer, as all infinities are hard to count — but hey, peanuts aren't proper nuts either. Cantor's discovery was the start of a true domino effect: the discovery of $\aleph_1$ led to the construction of an even bigger infinity $\aleph_2$, which in its own turn led to $\aleph_3$, and so on. Forever and ever, and ever.

Cantor's diagonal argument: in order to understand why there exist sets which are 'bigger' than $\mathbb{N}$ (and thus have 'more' than $\aleph_0$ elements), we will consider the set *D* containing all the real numbers between zero and 1. Note that in sharp contrast to the rational numbers (fractions), these numbers may thus have a decimal expansion that never stops and never repeats itself (a number like 0.981237237237... also has an expansion which does not stop, but it starts repeating itself at some point). For example, the number $\pi - 3 = 0.1415159265359...$ belongs to this set D. Now here is the ingenious trick that Cantor used: if this set D *were* to contain $\aleph_0$ elements in total, this would mean that we can make a list of all its elements, each of them indexed by a unique natural number $n \in \mathbb{N}$. Indeed, this is precisely what it means to have $\aleph_0$ elements: it means there is a one-to-one correspondence with the natural numbers, which implies that every number has a unique integer label. So in order to show that D has 'more' elements, Cantor essentially proved that there is no such pairing between the elements in D and the elements in $\mathbb{N}$, or that the elements in D can never be listed. And he did this by assuming that such a list *does* exist, and then deriving a contradiction. Before giving you the mathematical version, let's first consider a simple analogy. Suppose we have three people who answered three questions:

	Age	Hobby	Favourite dish
1.	<u>34</u>	hiking	pad thai noodles
2.	13	<u>skateboarding</u>	winegums
3.	86	complaining	<u>mashed potatoes</u>

If we now want to invent a *unique* profile, different from each of the ones above, a simple strategy is to go for someone who is not 34, and who does not like skateboarding nor mashed potatoes. This person will then automatically differ from person n, because we forced the answer to the nth question to be different. What Cantor did was exactly this, but then for an infinitely long list of questions, the answers to which are always numbers between 0 to 9, and where the 'people' have to be thought of as all the numbers in the set D.

So the thing is that if the elements of D can be ordered in a list, it would look as follows:

$$0 \leftrightarrow d_0 = 0.\underline{1}4159265359...$$

$$1 \leftrightarrow d_1 = 0.7\underline{1}828182845...$$

$$2 \leftrightarrow d_2 = 0.57\underline{7}21566490...$$

$$3 \leftrightarrow d_3 = 0.618\underline{0}3398874...$$

$$\vdots \qquad\qquad \vdots$$

To the left of the arrows are all the integers, and on the other side our alleged *countably infinite* list of decimal numbers between 0 and 1 (the answers to the infinitely many questions in the query). As you can see, *each* of these numbers has one digit after the decimal point which is underlined. These form the diagonal after which the argument is named, because what Cantor realised is that you can now use these digits to create a number which is not in the list (our unique profile from above). Indeed, if I write down the number $d_v = 0.2451...\in D$, then this number has the remarkable property that the first digit after the decimal point differs from the first digit of d_0, the second digit after the decimal point differs from the second digit of d_1 and so on (there are nine ways to pick these digits, as long as it differs from the one that is underlined). I can actually do this for all $n \in \mathbb{N}$, which means that d_v is different from *all* the numbers in the list. This can mean two things: *either* I simply forgot to add this number when I made the list, *or* it is simply impossible to compile such a list in the first place. Because I explicitly assumed at the start of the proof that I was listing *all* the elements — omitting none — one can only conclude that this list cannot exist.

In other words: the set D contains *more* than $\aleph_0$ elements. So this means that you need a new symbol to refer to 'the size' of this set, and that's exactly what $\aleph_1$ stands for.

So what Cantor basically proved[48] is that $\aleph_0^{\aleph_0} = \aleph_1$, and this is a new infinity — bigger than the infinity that we started from (the one 'counting' the natural numbers). If this all went over your head, don't be too hard on yourself: it's called the *diagonal* argument, it's not meant to be straight-forward.

Definition 20.

The Hilbert Hotel: *most people seem to be familiar with this place, as it often finds its way into popular references, but what is less known is that there exists a whole range of accommodation named after famous scientists.*

- **Cantor's Hostel:** *upon arrival at this place, you are shown a list with a complete description of every room in the hostel. Despite there being plenty of rooms, rumour has it that no one has ever been fully satisfied: every room seems to have at least one feature which doesn't agree with the guests' preferences.*

- **Darwin's Lodge:** *basically a chain of cabins in National Parks, often standing so near the lake that it seems like the buildings themselves recently crawled out of the water. Most customer reviews so far haven't been too positive, but we do believe there is room for evolution.*

---

48 Very technically speaking, it should be $\aleph_0^{\aleph_0} \geq \aleph_1$ (but then I need to elaborate on the Continuum Hypothesis, which leads us even deeper into the fascinating world of infinity). Let's just say there are too many inequalities in this world already, so I prefer $\aleph_0^{\aleph_0} = \aleph_1$ here.

- **Pythagoras' Backpackers:** *essentially one big room full of bunk beds, all of different sizes. Despite the fact that all the beds are too short, there is always a unique bed in which you will fit diagonally.*

- **Copernicus' Grand Royal:** *an upscale hotel which is very popular with the kind of people who are always dissatisfied with the view from the room, as this revolutionary hotel is essentially centred around the sun.*

- **Einstein's Motel:** *although specifically designed for people who are on the move, customers always seem to spend more time here than they intended — as if their clocks suddenly slowed down. It has been reported that the beds always look too short upon entering the room, but this never seems to be the case once you lie down in them.*

- **Heisenberg's Hotel:** *when you check in here, the staff can never tell you exactly where your room is. However, the longer it takes you to find your bed, the quicker you will fall asleep.*

- **Monty's Hall:** *you can get a decent bed at this notorious place, but this requires a bit of luck. When you arrive at Monty's Hall, you are shown three closed doors: one door leads to a luxurious bedroom, but if you choose one of the remaining two doors you will have to spend the night on a bed of nails in a mosquito infested room. There is some good news though: once you've made your choice, the staff open one of the other doors and offer you a chance to change your mind.*

3.3 Sir, someone broke the language!

All in all, it seems like your chances at winning the 'name the bigger number' competition are starting to look like underfed bait fish: bleak and slim. Mathematicians can counter both your finite and infinite submissions, with their powerful symbolic approach.

But what if we now do the opposite, and resort to full sentences again? How about we come up with the following entry for the competition: *the biggest number one can describe using at most 1,000 English words.*

Granted, it looks a bit more vague, but it definitely exists: the English alphabet consists of a *finite* number of letters, so amongst *all* the numbers that you can describe in 1,000 words there has to be a unique largest number. And yet, there is something really weird about this number. The problem is that the phrase 'one plus the biggest number one can describe using at most 1,000 English words' is bigger than my original entry, and yet I described it using *less* than 1,000 words.

> *'Sir, someone broke the English language!'*

The mildly disconcerting feeling you may be experiencing right now is not a brain fry, we have just bumped into a paradox. This one is known as the Berry paradox, although it was first published by the British philosopher and mathematician Bertrand Russell (1872-1970). One of the most interesting paradoxes in the history of mathematics was named after him: Russell's paradox. It concerns a peculiar set, denoted by means of the letter *R*, in that it contains all the sets which are not contained in itself.

> *'Say what?'*

Let us start from an easy example: in sharp contrast to the mathematically sound phrase that we have met in the previous chapter — that equation you can safely whisper into your dating partner's ears — there are also plenty of statements you'd better *not* utter on a first date. Things like 'This one here, on the left of the picture, is my favourite haemmorhoid,' or 'You somehow remind me of my mother, but you must be at least three times as heavy.' Now suppose we gather all these things into one big set *N*, the set of things you'd better not mention on a first date. So *N* is just a symbol, but it stands for a whole collection of things. And verifying whether something belongs to this collection is easy: just ask yourself whether you would mention it on a first date. In case the answer is 'Hell no, how stupid would that be?', you can safely conclude that this thing belongs to the set (so $t \in N$, where t denotes the thing).

The crucial observation to make here, is that 'the set containing all the things you would not mention on a first date' also has no chance of being voiced on that first date. Usual topics include work, favourite dishes and — two bottles of wine later — sexual preferences, but people rarely discuss the concept of an idealised collection of things that should not be mentioned. 'Hell no, how stupid would that be?' Well, exactly, but this therefore means that the set N belongs to itself, as it satisfies the requirements of belonging to the set N. Mathematicians usually express this as follows: $N \in N$. This may seem odd, a set containing a set, but it is perfectly acceptable. Let me give you another example: the set containing all the things that require at least six dimensions to be properly understood, contains the set containing all the tofu.

It is not so difficult to understand that not all sets share this property B of belonging to themselves: if we denote by the set of all round objects, then obviously $B \notin B$. This set contains pancakes (well, not mine, I always seem to end up with edible fractals), a distant star some 5,000 light-years from Earth with the unappealing name Kepler 11145123[49] and the table at which Lancelot and his peers were sitting, but not the set B itself. I mean, it is *a set* — a mathematical abstract concept — I wouldn't know what it looks like, let alone whether it is round.[50]

---

49 This star achieved the status of roundest natural object ever discovered in the universe, according to a study from November 2016. If you don't care about the word 'natural', you don't need to travel that far: just Google 'bubble butt'.

50 And before you ask: nope, not even the Venn diagrams representing these sets were round. Like I said: my teacher deducted points for that.

This means that we now have a property that sets *can* satisfy: some sets belong to themselves, others do not. This is the basic idea behind Russell's set **R**. Once again, this is just a symbol: it stands for the collection of all the sets which do *not* belong to themselves. So for instance: **B** ∈ **R** (as we have just established that **B** does not belong to itself), but **N** ∉ **R** (since **N** does belong to itself). Now, sets are like exclusive gentlemen's clubs: it all boils down to membership. So we could ask ourselves the following question: does **R** belong to Russell's set **R**? There are only two options here: yes or no. Unless you happen to be Indian: in that case the options are yes and the other yes. In symbols: **R** ∈ **R** or **R** ∉ **R**? Let us investigate which one it should be.

(i) Suppose the answer is *yes*, which means that **R** ∈ **R**. In other words: **R** satisfies the criterion to belong to the Russell set (the collection of all the sets which do not belong to themselves), so we can conclude that **R** is a set which does not contain itself. Using symbols: **R** ∉ **R**. But this leads to a contradiction, as we just started from the opposite assumption. Assuming that **R** ∈ **R**, we can conclude that **R** ∉ **R**.

(ii) So clearly the answer must be no, the other option. This means that **R** does not belong to the Russell set. Using symbols: **R** ∉ **R**. However, from the definition of Russell's set, this very fact grants **R** membership of Russell's set (it is precisely the criterion). Which thus means that **R** ∈ **R**. Yet another contradiction.

Definition 21.

The barber paradox: *Russel's paradox is often illustrated in terms of an easier example: imagine a town with a single male barber. As a rule, every man keeps himself clean-shaven in this town: he either shaves himself, or he goes to the barber (this is an exclusive 'or' of course, the other one would be really weird in this context). So the barber is the person who shaves all those who do not shave them-*

selves.[51] *The question then becomes: who will shave the barber? If he doesn't shave himself, he has to go to the barber, and thus shave himself as he is the barber — boom! This version of Russell's paradox should not be confused with the hairdresser paradox, which says that no matter how well your haircut is done, as soon as you walk out of the salon it will look worse than when you walked in.*

So what exactly is going on here? Well, we have basically bumped into a paradox; a statement which apparently contradicts itself. Paradoxes in itself are pretty harmless — except for that one guy who decided to *literally* rack his brains on one — but mathematicians do have to approach them carefully: they often reveal errors in definitions which were thought to be rigorous, and sometimes cause axioms of mathematics and logic to be re-examined. The paradox we have just described, discovered by Bertrand Russell in 1901, did just that. It shook the very foundations of mathematics and led to a large body of meticulous work (Russell's type theory and Zermelo's set theory), dedicated to recovering from this nearly fatal blow.

Definition 22.

Dotting the i's and crossing the t's: *an English proverb, which means that one has to be thorough and meticulous, or that all details — even the most minor ones — have to be taken care of. Paradoxically enough, the person who came up with this proverb clearly did not do this, since j's are dotted and f's are crossed as well.*

Like Berry's paradox, also Russell's paradox essentially revolves around the notion of a self-referential statement. It is a mathematical version of the liar's paradox, which in its simplest form reads '*this sentence is false*'.

---

51 I have always wondered why there is no Marvel superhero character called 'the Dutch Barberman': roaming the streets with his razor blade, ready to shave the world. We even have a main character for the movie: Sean Connery.

Definition 23.

Self-reference: *the property of a statement that is a statement about itself. See also: self-reference.*

Self-referential statements lead not only to paradoxes, sometimes they also induce infinite loops — yet another mechanism to mess with our minds. A nice graphical example is Escher's lithograph Drawing Hands from 1948, but the best illustration is edible: the Kit Kat chocolate bar. According to a BBC documentary from 2015, the middle layer of a Kit Kat is made from ground-up Kit Kat bars. Culinary philosophers were obviously dumbfounded, because it made them wonder what the first ever Kit Kat was made from. Quantum physicists are convinced that this actually proves that Kit Kat's Kat has to be Schrödinger's Cat: the first ever bar simultaneously did and did not have a middle layer, which solves the mystery.

Even my puberty was tainted by a self-referential loop, interfering with my social life. Whenever I had to ask permission to go out, I got stuck in one of the following dialogues:

'Mom, can I go to the party on Friday?' 'Go ask your dad.'

'Dad, can I go to the party on Friday?' 'Go ask your mom.'

Definition 24.

Puberty: *the phase in my life marked by the arrival of acne (turning my 15-year-old forehead into a rugged braille billboard displaying boldface headlines to the visually paired and impaired part of the population), a suspiciously large paper tissue consumption (salty souvenirs from the strolls along masturbation lane), and that disconcerting feeling once I realised that the moaning sound I could sometimes hear from the master bedroom when I went to the toilet was nothing but the auditory by-product of my parents doing their imitation of the bulgy beast with the two backs.*

The notion also lies behind what I personally believe to be the greatest *tour de force* in mathematics: in 1931, the Austrian mathematician Kurt Gödel (1906-1978) constructed a very peculiar self-referential statement in mathematics, which thus says something about mathematics itself.[52] Suffice it to say that the description of this statement in full detail lies beyond the scope of this book, but it is definitely worth pointing out that in doing so, Gödel fed the snake its own tail. Ouroboros may be shaped like a zero, but this surely wasn't nothing:[53] it turned out to be a tail-biting erhm, ... snake preview of his main achievement, as Gödel went one step (read: a giant leap) further and used this statement to prove his famous 'incompleteness theorem'.

Up until the day this theorem was proved, mathematicians were convinced that for *any statement* in maths, there are only two options (once again, no in betweens). Either it is possible to mathematically prove that the statement is true, or it can be shown that the statement is *false* (so that the negation of the statement is true). It is pretty obvious that a statement must be true or false — that's why I find maths so attractive, as there are no half-truths — but that is *not* what this is all about. The subtle point is the following: how does one *arrive* at this conclusion? What mathematicians thought (even hoped) was that 'a mathematical proof' would *always* be available to use as a criterion. For example, if I claim that the product of two odd numbers is even, one can (quite easily) *prove* that this statement is false. So this means that the opposite statement is true: the product of two odd numbers is not necessarily even (as a matter of fact, it will always be odd).

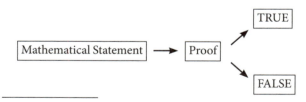

---

52   This is a bit like looking up the word 'dictionary' in a dictionary.

53   Violation of the well-known self-reference from your writing course: *never do no double negatives* (a so-called fumblerule).

What Gödel proved, is that this dichotomy of statements is wrong! He showed that *mathematical truth* and *provability* are not the same: truth is about the way things are, and proof is about what we can *know* for sure to be true.

Gödel's theorem shows us that there are mathematical claims that are true, but whose truth we cannot establish by a mathematical proof. These claims are bricks in the Mathematical Palace of Truth, but we can't prove it — they are forever beyond our grasp. It is perfectly normal if you have to read this sentence a few times: even I still struggle with the implications of this truly fundamental insight. Think of rainbow soup with meatballs, served by unicorns: we know it *has* to be good (it is soup! made from rainbows! with meatballs! served by unicorns!), but we just cannot *prove* it. This might be a bit of a stretch, but Gödel's remarkable result somehow makes me think of that wonderful thing called female intuition: women can sometimes say things that turn out to be true, but for which there is no explanation. As if they had access to the Database of Truth, without having to log in. I guess this is one of the reasons why maths is referred to as the *Queen* of the Sciences, rather than the *King*: there are truths that we just cannot explain. This result sounds massively interesting from a philosophical point of view, but it conveys a cruel message: there are statements in mathematics about something as elementary as the natural numbers ℕ for instance — which *have* to be true, but which *cannot be proved* using the standard axioms and laws of logic. Not because our mathematical knowledge is insufficient at this point in time, or because we have not been trying hard enough. It cannot be proved, for Gödel has proved it so.

'So how do you know these statements are true then?'

That is a very good question indeed, and this is where the abstraction level is moving up one notch. The thing is that he managed to cook up a formula which encodes the statement *'this formula cannot be proved'*. A true recipe for disaster, this formal version of the liar's paradox. What made Gödel's argument so ingenious is that he invented some sort of scheme to translate this simple sentence into a mathematical formula (a string of numbers and symbols). 'Like a code, you mean?' Yes, in a sense. But a difficult one, not your average boy scout cipher.

Once he had turned that devilish self-referential phrase into a formula, Gödel could treat it like a standard mathematical statement: 'Hello Mr Formula, can I prove you?' And this is where it all went wrong: suppose you *are* able to prove this formula, then the statement   suddenly automatically becomes false. Indeed, remember that the formula says 'I cannot be proved' (in coded language), and you just proved it. But here's the catch: saying that the formula is false means that its contents are not true (du-uh) — and in mathematics this means that the opposite statement is true. The negation of 'I cannot be proved' becomes 'I *can* be proved', and this is even worse. You have now found a *false* mathematical statement which can be proved: this is like claiming that you can *prove* that 1 is equal to 0. And apart from broken coffee machines, a chronic chalk allergy and negative epsilons,[54] this is the mathematician's very worst nightmare.

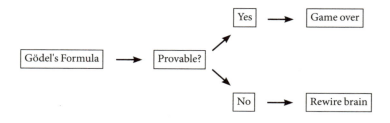

However, there is a way out of course: this whole mess started because we assumed that Gödel's formula could indeed be proved (which turned it into a false statement). So if you want to avoid

---

54   Many theorems start as follows: 'Let $\epsilon > 0$ *be a positive real number*'. A standard joke amongst mathematicians goes as follows: 'Let $\epsilon < 0$. *<insert tumbleweed here>*

mathematical proofs for false statements (and yes, we want to avoid them), you have to go for the other option: you will have to accept that you cannot prove the formula, *which is what it was trying to tell you from the very beginning.* So we have saved maths, but we have to live with the following inconvenient truth: Gödel's formula is an example of a mathematical statement that is true, but we just cannot prove it.

Post-Gödelian eyecandy

3.4 You are smarter than a pirate

Okay, so far we have learnt what sets are and we have seen how they shook the very foundations of the mathematical construct. But going back to our original question: exactly how do we get from 'sets' to 'spaces' then? In other words, what does it mean when mathematicians say that 'extra structure' can be added to a set? Think of sets as online dating sites: membership is one thing, but in order to make it interesting you have to add more information. Anything that sets you apart from the other people sitting around that pond, trying to avoid (or attract) the odd duck.

Including a profile picture which often fails to do what it was designed for — to exude an air of casualness — as it usually seems to be the result of a selection strategy so severe that it makes Google's hiring process look like a kindergarten admission test.

For instance, the set $\mathbb{Z} = \{\ldots, -2, -1, 0, 1, 2, \ldots\}$ containing the positive and negative integers already has *more* structure than the set we usually call 'the pile of dirty dishes in our kitchen'. You are probably so used to it that it does not strike you as a peculiarity, but the fact that integers can be added, subtracted and multiplied with each other distinguishes a set like $\mathbb{Z}$ from more mundane sets, such as those bloody dishes, or your record collection. If only that set of dirty dishes came with a unary operation like the negation, sending any number a in $\mathbb{Z}$ to its opposite value $(-a)$, and a binary operation like the addition, which sends two numbers a and b to their sum $a + b$. Doing the dishes would be as simple as

$$\text{dirty plate} + 2(-(\text{dirty plate})) = \text{clean plate}.$$

Now I know that most people hate algebra, but I am talking about serious spaghetti sauce leftovers here, at least three weeks old (Fifty Shades of Green).

Also note that the set $\mathbb{Z}$, containing positive *and* negative integers, carries more structure than the set $\mathbb{N}$. You can still add positive integers, but the operation $a \mapsto (-a)$ is no longer defined for a in $\mathbb{N}$, as there are no negative numbers in $\mathbb{N}$. In the same vein, the set $\mathbb{Q}$ containing all the fractions carries more structure than the set $\mathbb{Z}$: the former has the additional property that two elements can be divided (well, as long as you don't divide by zero[55]), whereas the quotient of two arbitrary integers is not always an element of $\mathbb{Z}$. This is the reason why mathematicians use different words to describe sets with different structures. It then allows them to somehow 'forget' the particular structured set they started with, and to focus on the more abstract concept associated with it.

---

55   I am telling you, the puppy will die!

For example: $\mathbb{N}$ is an example of a monoid, $\mathbb{Z}$ serves as a prime example of a ring and $\mathbb{Q}$ is a so-called field. This means that anything you can say about monoids, groups or fields in general will thus automatically apply to the sets $\mathbb{N}$, $\mathbb{Z}$ or $\mathbb{Q}$ in particular. Precisely this philosophy lies behind one of the more modern descriptions of mathematics as 'the study of patterns and structures': start from a set you like, and turn its properties into the defining characteristics for an abstract structure (a space) that you can now study on a meta-level.

Unary and binary operations are but two easy examples of how you can impose extra structure on a set, but there are many mechanisms, leading to different kinds of (mathematical) spaces. Without going into too much detail here, one can think of these 'mechanisms' as 'things one can do with the elements in the set': performing numerical operations (like adding and multiplying), comparing elements, measuring them in some sense, or even stretching and bending its members. This gives rise to a huge collection of different structures, like countries on an abstract world map: some of them are neighbours, but some belong to completely different continents and have their own language.[56] I bet there are still plenty of pristine islands out there, waiting to be discovered. Pirates, set sail! Or better: pirates, wait until you have read the end of this chapter, and set sail!

One can, for instance, add a metric to a set, which then turns it — hold on, this is going to blow your mind — into 'a metric space'. Or you can be slightly more demanding and go for an inner product, turning your set into an 'inner product space'. One of the most crucial concepts in mathematics is that of 'a vector space', essentially a set furnished with two compatible binary operations. And if that is too straightforward for you, you can almost literally deform these spaces, turning them into so-called 'manifolds'. You can even go a bit more exotic and add a topology to a set, which

---

56 Using an actual space as a model for more abstract spaces, does that qualify as a meta-metaphor?

leads to the study of 'topological spaces'. Or how about adding coin-operated showers which run out of hot water as soon as your head is shampooed, mosquitoes, a bunch of Dutch people who arrived in what can safely be described as a cross between a trailer truck and a small school building, a handful of Germans and a little shop selling disposable gas cartridges, knives and alcohol. This turns your set into 'a camp site in the south of France'.

Definition 25.

Mosquitoes: *no matter how high 'pure mathematics' scores on your list of 'least useful things in the world' — presumably in close contention with those little green pieces of plastic grass in a box of takeaway sushi — it does not even come close to the outright winner on the official list: the mosquito. Not only does this flying mugger quite literally suck, I also believe that its highpitched buzzing qualifies as one of the most annoying noises in the world.*

It seems that a mosquito sucks around five millionths of blood per bite, which boils down to 0.005 millilitres. Since an adult person typically carries five litres of blood around, this means that one million moquitoes could kill you overnight. A fitting end for a pessimist: life sucked.

Now, let us assume that you are facing 70 years of being bitten — I suppose baby blood is to the flying bastards as unripe avocados are to grown-up people: not worth consuming yet — with an average of 100 bites per year. This is a pessimistically awful lot if you ask me (I bet you had emotional break-ups that were less draining), and most people will resort to mosquito repellent and nets after a while anyway, but let us take this number to be on the safe side. If we then multiply everything together, we are led to a grand total of 35 litres of blood over the span of a lifetime (70 × 100 × 0.005).

Here's an idea: can't we just organise some sort of Annual Donation Event, where visitors can have 500 millilitres of their blood drawn and then send a signed bag to Mosquitoville? Consider it like yet another sort of bloody tax.

And just to make sure that the itchiness and the maddening noise are covered too, I even suggest forcing donors to wear a t-shirt made out of wool, and a set of headphones playing a mix of nails scratching on a blackboard, squeaking Styrofoam and French news readers covering the Tour de France.

DO YOU LIKE TRUE BLOOD?

NAH, I PREFER THE ITCHY AND SCRATCHY SHOW

Definition 26.

Mosquito net: *a device that looks like a fine, see-through mesh construction which can be hung over the bed (from the ceiling or a frame) or built into a tent, and works more or less like a fish trap for small, biting insects spreading tropical diseases. Unless the mosquito net is broken, there will always be at least one tiny hole through which insects will enter the trap but cannot leave. Most mosquito nets are impregnated with some kind of repellent to ensure that this hole can easily be found. For optimal results, people sleep under these nets having applied mosquito repellent to their body: this drastically increases the chances of being bitten at the desired position.*

Definition 27.

Mosquito repellent: *a milky substance applied to human skin, in order to attract mosquitoes and encourage them to bite at particular places of the body (this can be anything from a finger or the tip of an elbow to a spot on the ankle). This is accomplished by covering the body with this substance almost completely, thereby deliberately avoiding those spots meant to be bitten.*

Sorry, I got somewhat carried away. Mathematical spaces, that's where we were. In order to illustrate the difference between a mere set and a space, I will consider an example and focus on the so-called metric spaces, which are obtained by adding a 'distance function' (sometimes called a 'metric') to a set. This is essentially a means to calculate 'how far' two objects in a set are separated from each other, which means that they behave like rulers. 'What, so they have their heads printed on stamps and erect unnecessarily big buildings to underline their grandeur?' No, not the people, I mean the instruments: given two elements in a set, a metric is the tool with which you can determine the distance between them.

An easy but interesting example is the following: let us denote by **P** the set containing all the points *p* in the plane which are described by coordinates of the form (x, y), where x and y are real numbers (think of these as dots in a grid, just like in the previous chapter). We can then upgrade **P** from a mere set of points to a (metric) space by adding a metric. For that purpose, we thus need a way to calculate the distance between two points p_1 and p_2. And you know what they say, 'where there's a will, there's a formula'. As a matter of fact, this formula is given by

$$\text{distance between } p_1 \text{ and } p_2 = \sqrt{(x_1 - x_2)^2 + (y_1 - y_2)^2}$$

Chances are that you recognise this formula, as it is nothing but Pythagoras' theorem — be it in a slightly different version than the one you banned from your maths memory ($A^2 + B^2 = C^2$).

For instance, for a point p_1 with coordinates (6, 1) and the point p_2 with coordinates (3, 5) we get $\sqrt{(6-3)^2 + (1-5)^2} = \sqrt{9+16} = 5$. This is the distance between both points, as the crow flies. Or any other thing that flies. Well, apart from actual flies: not only would that be linguistically poor ('as the fly flies'), the stupid creature would probably get stuck between the window and the open veranda door.

Regardless of whether you recognised the distance function above or not, it is by no means the only possible metric on the set **P**. As a matter of fact, you can make your distance function as crazy as you want, as long as it satisfies the following four requirements:

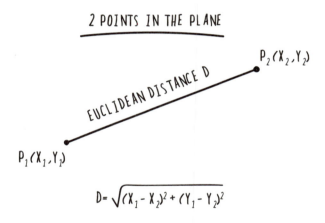

2 POINTS IN THE PLANE

$P_2 (X_2, Y_2)$

EUCLIDEAN DISTANCE D

$P_1 (X_1, Y_1)$

$$D = \sqrt{(X_1 - X_2)^2 + (Y_1 - Y_2)^2}$$

- *The distance between two different elements in a set is always a positive number.* If someone ever tries to convince you of the opposite, politely nodding and backing away slowly should do the trick. Calling the police is optional.

- *The distance from an element to itself is zero, and vice versa.* Just imagine how tiring your life would be if you had to cover an actual distance to go from the place where you are standing to that very place where you are standing.

- *The distance from an element p_1 to an element p_2 is the same as the one from p_2 to p_1.* This requirement is referred to as 'the symmetry of the distance function'. It can also be found in Captain Obvious's list of well-known facts, right under 'soluble soup bowls defy their purpose'.

- *Going from p_1 to p_2 first, and then from p_2 to p_3 can never be shorter than going from p_1 to p_3 directly.* This requirement is also known as the 'triangle inequality', and it is common knowledge amongst housewives: it essentially says that stopping at the pub when running an errand can never be a short-cut.

Just like the elementary operations in $\mathbb{Z}$, we are so used to working with 'distances' between points that we often forget that they are special. This makes thinking about spaces without a metric, or spaces with a different one, counterintuitive. The reason for this is evolutionary in nature, or nurture, as we all grew up in a Euclidean universe (which is the one where we use Pythagoras' theorem to calculate distances). Well, one day this may turn out to be completely wrong, as cosmologists are still trying to understand the shape of our universe. This may come in handy the day someone finds out that the meaning of life is to put our cosmos into the appropriate black hole of a giant shape sorting box. But let's say it *feels* Euclidean.

No surprise there of course, because from an early age onwards we all became acquainted with angles — usually in the form of a sharp corner interfering with our adventures in Furniture Land — and the difference between 'far and near' — spoon-fed by our mother, chartering daily direct Air Mash flights from Jarville to Mouth City. But this notion of distance that we have all been experiencing since birth is a very particular example. It is called the 'Euclidean distance', named after the Greek mathematician Euclid whom we met earlier in the context of the fifth axiom.[57] He is mostly known as the author of *The Elements* (written around 300 BC), a mathematical treatise of no less than 13 books on geometry and algebra which counts as one of the most influential works in the history of mathematics,together with Isaac Newton's *Philosophiae Naturalis Principia Mathematica* (from 1687) and the somewhat lesser-known *Interesting Daytrips Outside the Ivory Tower*. The legacy of Euclid is still very much alive and kicking: every year high school teachers all over the planet are trying to launch young people into that parallel geometric universe where lines, circles and polygons seem to live in perfect harmony.[58]

We are actually *so* used to working with the Euclidean distance that we sometimes tend to use it when it is not appropriate. Being a keen traveller, I sometimes overhear the following conversation from an adjacent seat on the aeroplane:

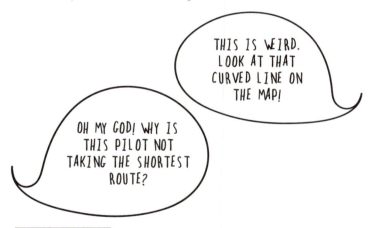

---

57   For all the kids out there, there's a lesson to be learnt here: you can't expect people to name a distance function after you if you start throwing people overboard — and 'but they're saying not all numbers are rational' is not a valid excuse.

58   They only *seem* to do so, as Edwin Abbott Abbott explains in his satirical novella *Flatland: A Romance of Many Dimensions* — another classic in the maths section. It was first published in 1884, and combines sharp criticism of the hierarchy of Victorian society with an attempt to illustrate the concept of higher dimensions to a general audience.

No (shortest) way!

First of all, I would like to point out that sincerely believing that you can outsmart someone who spent a few years of his or her life studying advanced mathematics, physics, meteorology, aviation technology and aerodynamics — also mastering the art of not spilling drinks during turbulence — crosses the lines of smugness *so* badly that rescue helicopters will have to be sent out for you (ones that *will* follow that curved line), since only God knows the unchartered territory where you have ended up.

Secondly, I do forgive you, as this principally proves my point: the notion of a Euclidean distance is etched deeply into our minds. Ever since the start of our evolution from the bouncy amoebas in the primordial soup we once were, that vast neural network which we call 'our brain' has adapted itself to a life in three dimensions, governed by the classical laws of physics — the ones describing falling pieces of fruit, swinging pendulums and dutiful planets.[59]

---

59   Your brain also learnt to filter out other things: you can *always* see your nose — it is constantly visible in your line of sight — it's just that your brain learnt to ignore it. My life would be a bit more bearable if my brain could do the same trick with people carrying a selfiestick.

One could even argue that this is the reason why topics such as quantum mechanics, general relativity — and the fact that Gary Numan is older than Gary Oldman — are so hard to digest: we are quite literally not made for it. Elementary particles and black holes are too small or too far away: from an evolutionary point of view there was no need whatsoever to develop some kind of 'intuition' regarding these matters. One could almost argue that studying (higher) mathematics is part of a process which aims at rewiring our brains, in order to better understand the universe. And also to seduce the sapiosexuals.

As part of this rewiring process, let me explain how to find the shortest distance on a spherical object like the Earth:

- Take an orange.

- Grab a pen.

- Draw two arbitray dots on the orange — feel free to add a smiley face if that makes you happy.

- Get a rubber band.

- Place it over the orange in such a way that it goes through both dots. There should be a unique position where the rubber band stays in place, and that's when 'it cuts the orange in half' (as if it were the equator). Any other position is not stable, in the sense that the rubber band will start moving and slide off the orange.

- Connect these dots, drawing a segment along the rubber band. Unless you started with diagonally opposed points, you have to choose the shortest segment here (measured along the band).

- High-five yourself. You've just found the shortest distance between two dots on a sphere (the orange). In mathematical lingo: you have drawn a *geodesic* on a piece of fruit.

While we are at it, let's do some spherical geometry (one of the counterparts to classical Euclidean geometry, in which the Playfair axiom is replaced by one of the other alternatives). In Euclidean geometry, where parallel lines through dots are unique, geodesics also exist: they are nothing but straight lines. So, these great circles on a sphere — the position of the rubber bands on the orange — are the 'straight lines' in spherical geometry. Just like metrosexual guys, it's not because they do not look very straight that they aren't: for an ant moving on the orange, this geodesic *is* a straight line. Now for any great circle and a point not on that circle, it is impossible to draw a second great circle which goes through that point but does not intersect the first great circle. If you draw a few great circles through that point, you will see that they always intersect. That is exactly what we expected from the first alternative for the Playfair axiom.

It is only when we want to recreate spherical geometry on a piece of paper — or a computer screen — that this shortest line between those points on the sphere (the segment of the rubber band) becomes a 'curved' line (with reference to the Euclidean metric). You can easily check this for yourself, by peeling the orange and flattening it out.[60] This flattening process will turn the 'straight line' on the orange into a curved line on the table. Not even if you do it very gently: you simply cannot represent the surface of a sphere on a (flat) piece of paper, unless you are willing to accept that distortions will appear. This fact was proved by the German genius Carl Friedrich Gauss (1777-1855), in a famous theorem called 'the Theorema Egregium', which says that every map of our world is wrong (because of the distortions, that is, not because of the funny colours they always use). You can also see it the other way round: the only way to wrap that soccer ball you bought for your nephew's birthday, is to crumple the wrapping paper.

---

60   If you end up with a stained shirt, you probably flattened the orange. Sorry, I meant the peel of course.

Look, not even the Japanese know how to do this — and they certainly have taken 'paper wrapping' to a level where only fire-fighters can reach it with their telescopic ladder — so it is safe to conclude that it is genuinely impossible.

Despite the fact that I was making fun of people mistaking curved lines on a map for detours earlier on, I do have to say that this is still a gazillion times better than what pirates made of it.

'Pirates, you say?'

Yes, pirates. The unruly rulers of the seven seas, the obnoxious oligarchs of the ocean. The bold and bearded boat-bothering brotherhood. Nothing but a gang of handicapped old men on a ship, that is what I think. An eye-patch, a wooden leg and a prosthetic hook[61]; it almost sounds like a kit to play an impaired version of rock-scissors-paper. They may have caught many women, collected plenty of gold and cracked lots of booze, but if there is one thing they did not capture it surely was the definition of the distance function on a sphere. Just take a look at a pirate's treasure map. I do understand the part where they are looking for the 'X' — that sounds perfectly normal to a mathematician — but for God's sake, which metric are they using?

---

61 Judging from their unkempt facial hair I think it would have made more sense to furnish them with a razor blade.

That can hardly be a geodesic, right?

Recommended listening

Artist	Song title
Bane	Count Me Out
Sick of It All	Who Sets The Rules
Beastie Boys	Remote Control
Raum Kingdom	Cross Reference
Pennywise	Set Me free
Clark	There's a Distance in You
The xx	Infinity
Bloc Party	Helicopter
The Dillinger Escape Plan	Calculating Infinity
Moderat	Out of Sight

4

Space To Kiss

*Any man who can drive safely while kissing a pretty girl
is simply not giving the kiss the attention it deserves.*

You should really read this chapter if ...

- you thought you were the only one with kissing problems.

- you are in the mood for a shorter chapter.[62]

- you want to use a set of snooker balls in an experiment,
  hereby illustrating a famous argument between Newton
  and one of his peers.

- you really want to know the size of your personal bubble
  space.

Like I said in the introductory chapter, the physicist's definition
of space — the enormous stage on which all things exist, events
happen and history unfolds — more or less coincides with my gut
feeling of what it is. A feeling which occasionally rears its ugly
head, like when there is either *too much* space or *not enough space*.
I have to admit, I do not really know how it feels to have too much
space at my disposal — I once went to an outdoor fundraising
event for agoraphobics but none of them showed up — but there
are quite a few situations in which I feel like there is not enough.
Situations in which my personal bubble looks like a piece of fruit,
being prodded by a picky customer testing its ripeness.

---

62 If you feel like pointing out that this depends on the metric being used here:
    <bonus level unlocked>.

With that same self-proclaimed authority some people try to emanate when tasting the wine in a restaurant.

Situations such as me boarding an aircraft for instance. This has nothing to do with fear of flying by the way, as my faith in pilots is similar to my faith in chiropractors: if not utterly blind, then at least cross-eyed, nearsighted and suffering from cataract. As long as they do not break anything vital, I am fine with whatever they are doing.

My problem with flying can be traced back to 2003: I was 24 years old, and getting paid to travel to the other side of the world for a mathematical conference. My sense of self-accomplishment was still in its infancy at that time, but flying to Sydney to lecture about things that I had invented myself surely boosted its growth. As soon as I had boarded my flight though, these feelings of pride and contentment had to make room for thitherto[63] unknown unease, and my neighbours: I had absolutely no idea how I was supposed to survive 24 hours on an aeroplane, wedged between two horizontally tall passengers, who had already claimed both my armrests and my prospects of a good night's sleep. I felt exactly like how they pronounced my last name later that week, when they announced me as a speaker: *elbowed*.

Everyday Isomorphisms: *An aircraft interior designer providing three people in a row with four armrests only is like a waiter setting the table by putting only one piece of silverware between two plates.*

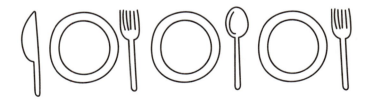

---

63   This word does sound very archaic, but words should not be treated like fine
      china, stowed away in the cabinet and rarely used.

In hindsight, I now understand why the ladies behind the counter were brazenly checking me out before checking me in — boosting my masculine confidence with that sensuous smile which transcends professionalism. Back then, it simply did not occur to me that they were just eyeing me up to estimate my body weight: they were probably just waiting for someone who was slim enough to fit between two gravitationally challenged people, yet not assertive enough to make complaints about it.

Since then, the same thing has happened to me quite a few times. Every once in a while I do get lucky though: I am seldom overlooked by the cabin attendant offering what seems like randomly selected passengers a chance to change seats when the flight is not overbooked. Mind you, this is actually a typical example of the Law of Conservation of Misery. Each time a sensible stewardess allows me to change my seat, I am somehow bound to end up next to an embodiment of the five remaining degrees of exasperation: an ADHD child washing down energy bars with cans of Red Bull, a wardrobe-sized Finnish guy mistaking the 12-hour flight for a vodka tasting, a compulsive sleeper whose excessively drooling head switches between neighbouring shoulders like a dribbling metronome, a loquacious lunatic trying to convince me that we live in a universe which has 17 dimensions, or a French guy.

Definition 28.

French: *an artificial language devised by a team of linguists and foodies, with the sole purpose of increasing the attractiveness of expensive meals. This is the reason why an hors d'oeuvre sounds better than a mere starter. Other well-known mechanisms to achieve this goal include the usage of extremely long and vaguely poetic descriptions for food items — often involving seasonal words and nouns referring to pieces of furniture — and the addition of totally irrelevant facts.*

This leads to typical menu entries such as 'escargots on a squeaky bed of wet monsoon vegetables, sprinkled with freshly ground purplish pepper from a traditionally hand-shaped pepper mill which was found in the tomb of a young Pharaoh who died from an onion overdose'.

In a sense, being wedged in between only two other people in a row is still better than the mathematically correct answer to the question: what is the optimal way to fill up the economy class? Assuming that we are all granted the same amount of space around us (read: the personal bubble radius is a constant), this leads us to the *kissing number problem*.

Please do not confuse this with the number of kissing problems. I bet at some point during our lives we were all wondering in which direction to tilt our face when the moment was finally there, speculating how far to open our mouth and trying to synchronise the angular velocity of our tongue with our partner's. Apart from the question 'Why is French kissing called French kissing?' — I stopped pondering, they can have it as long as they give us Belgians our fries back — my biggest kissing problem was undoubtedly that piece of barbed wire my dentist decided to adhere to my teeth. Dating was not just the proverbial blood, sweat and tears in my case: I am still being haunted in my dreams by girls with bleeding gums.

The mathematical 'kissing number problem' seeks to know the maximal number of non-overlapping equal-sized balls that can touch a given ball. Yes, that is a mouthful, but then again: they did not call it the kissing number for nothing. Note that other names for this problem are 'the Newton number' (after the originator of the problem, also known for his rather striking contact with a piece of fruit) and 'the contact number' (which I do find a bit weird: you can call me old-fashioned, but I grew up in an age where 'contact' was reserved for a few dates later).

On a line (think of an airline), the kissing number is equal to 2. Remember the last time you flew somewhere: your personal bubble in the middle, kissed from both sides — like the winner of a road bicycle race at the end of the medal ceremony. I'm going on a bit of a sidetrack here, but I have always found these two lousy kisses small comfort for the athlete in question. I mean, as soon as the winner of the race crosses the finishing line, he is immediately surrounded by a pack of journalists, who start firing border-line ridiculous questions at the unfortunate road cyclist. 'What was your plan when you started the race today?' Gosh, I don't know, Sherlock, maybe riding that bike as quickly as possible? In the meanwhile, the sportsman is desperately trying to catch his breath — what would you do if you'd just spent half a day on a saddle which looks like a padded rock pick? — reducing the inter-view to a verbal rapid-fire, interrupted by the exhausted athlete's mumblings. This apparent lack of rhetorical skills in the world of professional sports might be the reason why there are no disci-plines for mute people at the Paralympic Games: it is so common that it simply does not qualify as a proper handicap.

'[...] is simply not giving the kiss the attention it deserves'

In an aeroplane — luckily enough not on a plane, but let us all cross our fingers and hope that Ryanair's Michael O'Leary is not reading this book[64] — the kissing number is equal to 6. It is not too difficult to see why: the triangle in the picture below is equilateral, which means that it has three sides of equal length (given by the diameter of these circles) and three angles which are equally big. As the sum of the angles in a triangle is equal to 180 degrees (unless if you bought this book in a non-Euclidean universe), this thus implies that each of these angles is equal to 60 degrees. In particular, the angles meeting each other in the centre add up to 360 degrees and this happens to be a full circle. So you cannot add an extra circle kissing the one in the centre. This is the reason why snooker players cannot fit more than six balls around any other ball at the start of a game, when they fill up that wooden triangle.

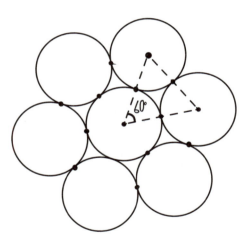

Kissing number 6 in two dimensions (in a plane).

---

64   I do charge an additional fee of 50 euro if you want to read these pages in the right order, Mr O'Leary.

Definition 29.

Snooker: *the perfect game for lazy people: it is supposed to be played on a bed lined with cushions, a player's goal is to make his break as big as possible and after that he can then take a rest.*

Now let us take this kissing game one level further: imagine you take the white ball from the snooker game and you want to know how many red balls you can make the white one kiss in space. Wait, before you start launching snooker balls into our cosmos, I mean 'in space' as in 'you are allowed to make use of the third dimension'. In other words, you are allowed to lift the balls off the table this time, but they must all still have contact with the white ball in the centre. Can you then guess what the kissing number will be in three dimensions?

To visualise this, you could start from the configuration in two dimensions and add balls from above. Next, you turn the result upside down and repeat the procedure. Because of this symmetry, the conclusion would be that the kissing number in three dimensions is equal to six plus twice the number of balls you can add from above. If you were to try this explicitly, you will notice that you can add three balls on top of the seven balls already on the table (the six red ones around the white one). There is a total of six 'valleys' so to speak (formed by two red neighbours and the white ball), but you can only park a red ball in every other one of them.

THIS BALL IN
THE MIDDLE IS
WHITE

(THE OTHER
BALLS ARE RED)

THIS BALL
LIES ON
TOP

THIS BALL
LIES ON
TOP

✱ THE POSITION OF THIS
BALL IS LEFT AS AN
EXERCISE FOR THE READER

So your final answer is 12 then? A mathematician would definitely not be convinced yet. Because what you have just done, is to *empirically prove* that the kissing number is *at least* 12 (in math lingo: 12 is a 'lower bound'). From an experimental point of view this sounds very satisfactory, but mathematicians need more than that to be fully satisfied. A picture is not a proof, and neither is an experiment. Monkey see, monkey do, but monkey still not convinced.

So how do you know for sure that you can't park an *extra* red ball around the white one? I will actually play the devil's advocate here and give you two arguments supporting the claim that maybe you could.

Argument 1: First of all, this configuration of 12 red snooker balls kissing the white one in three dimensions is not 'rigid'. This stands in sharp contrast to the situation in two dimensions. By that I mean the following: there is a little bit of space between the red balls, in the sense that you can always isolate a red ball so that it only kisses the white ball at the centre, but not its red neighbours. I say 'a little bit of space', but there is actually enough space to move the red balls around and this allows you to change the relative position of any two balls of your choice by just rolling them around over the central white one. Just to make sure, this means that these red balls keep kissing the white ball at all times — there's no need to break the contact between the red ball and the white one. In two dimensions, this is clearly impossible: you can change the relative position of two red circles, rolling one over the other, but this is not allowed as it violates the kissing property. In a sense, there is literally not enough space in two dimensions; each circle is locked into position by its neighbours. We already met this phenomenon when we talked about moshing versus breakdancing: sometimes things get easier when the number of dimensions increases, because this allows for more wiggle room (this does lead to curious consequences though, see the chapter on slicing and stacking). Because of this wiggle room, you may ask yourself — just like David Gregory, a famous astronomer, a gifted mathematician and a contemporary of Isaac Newton, did in 1694 — whether it is possible to move the balls around in such a way that some space opens up for an extra red ball.

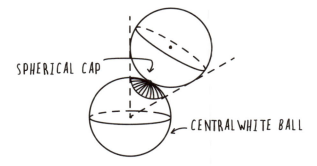

SPHERICAL CAP

CENTRAL WHITE BALL

Argument 2: Space opening up for one extra ball is even a bit modest: with some high school calculus and geometry, one can prove that the kissing number in three dimensions can be *at most* 14 (in maths lingo: 14 is an upper bound). Without going into details about the calculations, I will give you the main idea behind the argument: imagine putting each of the red balls in a cone — like when people place a funnel around the head of a dog that is recovering from a healing injury, to prevent it from licking that other kind of red balls. If the apex of this cone sits at the centre of the white ball, just like in the picture above, it will cut a certain area from the white ball, which mathematicians call a *spherical cap* (think of the pope's skull cap). Now if the kissing number is equal to the number k, you will have k of these cones cutting off k non-overlapping spherical caps from the central white ball. This means that k times the area of this cap must still be less than the *total* area of the white sphere. Think of it this way: the *maximal* number of pizzas you can put on your kitchen table is the area of your table divided by the area of a pizza. More often than not, this maximal number cannot be reached — unless you start slicing some of the pizzas, which we will do in a later chapter, or you happen to have a very odd kitchen table — but it is definitely impossible to put more on that table. For tables (rectangles) and pizzas (circles), calculating the area is fairly simple, that's just standard household mathematics. For spheres and spherical caps this is slightly more involved, but it can be done using a formula expressed in terms of an integral — the curly symbol of doom. Now if you were to calculate the surface area of the white sphere (the area of the table so to speak) and you then divide it by the area of the spherical cap cut out by a red kissing ball (the area of a pizza), you would get a tad under 14.99. This means there is *just not enough* space for 15 red balls.

Integrals: amongst the many things nightmares are made of — pizzas with pineapple, Japanese high school girls whose hairdresser clearly took the curtained hairstyle a bit too literally,[65] or being born in a world made of Lego without shoes — integrals probably rank as one of the most scary ingredients.

CREEPY CHILD FACTOR   CREEPY CHILD FACTORY

There exist many different kinds of integrals (I suppose this is a bit like saying that there exists more than one kind of hairy spiders), but the most famous one is definitely the Riemann integral, the one you may have learnt about at school. It is named after the German mathematician Bernhard Riemann (1826-1866), who is actually a prime example of a mathematician who paved the way for practical applications through purely fundamental research. Apart from the concept of an integral, he also invented a kind of geometry, later dubbed 'Riemannian geometry' in his honour, which set the stage for the theory of general relativity. Because Einstein realised that physical space is

---

65   To be honest, I have never really understood Japan's fascination with little longhaired girls in scary movies. When it comes to horror hair, nothing beats mullets. And purple grannies.

not merely a metaphorical stage: it is a *curved* metaphorical stage. As with Rubens' women: mass induces curvature (Einstein's insights in three words), and it was Riemann who introduced the tools required to describe mathematically how curvy things are. Anyway, when it comes to integrals, it sometimes suffices to expose people suffering from maths phobia to an expression like

$$\int_0^\infty x^2 \frac{\cos \pi x^2}{\cosh \pi x} dx = \frac{1}{8\sqrt{2}} - \frac{1}{4\pi} \, ,$$

and chances are you will soon find yourself scrubbing barf from the floor. For me personally, though, integrals were my gateway into the world of higher mathematics. I must have been still at primary school when I got a book from my favourite uncle with tables of formulas for surface areas and volumes for geometrical figures. I diligently learnt them all by heart, until one day he spoke the following enigmatic words: 'Once you can integrate, you will be able to make your own formulas.'

If only I could stick to my New Year's resolutions as well as those words stuck to me: they basically guided me through my secondary school maths classes. Because I had absolutely no idea what he meant at that time, but I do remember that I was eagerly looking forward to the day I would finally understand his words. During my last year of secondary school it finally happened: I learnt to calculate integrals, reducing those tables of formulas to mere exercises. I still remember calling him after school that day: 'Dear Uncle, I no longer need your book: all I need now, is a piece of paper and a pencil.' In hindsight, I wish I had asked him for a porn magazine for my birthday that year — he is the kind of uncle who would have given me one. Just imagine his face after that phone call.

Getting back to the kissing number problem in three dimensions, the integral formula for the surface area of the cap seems to suggest that it could be as much as 14. Newton, on the other hand, was convinced the kissing number was equal to 12, whereas Gregory thought that — unlike hotel room numbers and aircraft seat maps — there was space for number 13. So which one is it?

Definition 30.

Triskaidekaphobia: *fear of the number 13 and avoidance of using it. For the readers suffering from dyscalculia: fear of the number 31 and avoidance of using it. For all the readers suffering from Helleno-logophobia: sorry.*

Note that in sharp contrast to universes, this problem did not just come out of nowhere. The kissing number problem allegedly arose when Newton and Gregory were discussing astronomical issues. One of the points they were debating was how many planets can revolve around the sun and — a phrase often used in the context of kissing — one thing led to another. Neither Newton nor Gregory ever knew who was right during their lifetime: despite many attempts and nearly finished arguments, a sound proof was not found until 1953, which is more than 250 years after the question was posed. It turned out that Sir Isaac Newton was right after all: the answer is indeed 12.

'So what lesson is in there for me,' I hear you asking, 'because this just sounds like a bunch of abstract nonsense without real-life applications.' I can think of at least one lesson we can draw from the kissing number problem and its rich history: until that glorious day we can place a mathematical proof under our windscreen wipers showing that there will indeed not be enough place for that one extra car along the side of the road, we might as well park our own vehicle just that tiny bit closer to the one in front of us. You never know.

But there are plenty of other applications once you realise that the notion of a kissing number is merely a special example of a so-called spherical code. You see, one of the main characteristics that sets mathematicians apart from other people is the following: the latter are happy when they find out there is a solution, the former are unhappy because this means there is no longer a problem. So what often happens in maths is that people try to *generalise* problems: looking at it from a different angle, transposing it to more dimensions or demanding a proof that fits in the margin of a book. And so once mathematicians figured out the kissing number, they tried to generalise this notion and it turned out that determining 'spherical codes' provided them with a whole new bunch of problems. These abstract objects are studied in geometry and coding theory, but they do have applications in real life. For example, the study of the microbiological structure of viruses — some of which, by the way, are kissing problems — is based on it.

A spherical code is defined as a particular configuration of n dots drawn on a sphere — formed according to a special set of rules. You can even do it on hyperspheres, but these mathematical monsters won't be defined until a later chapter ('Too Much Space'), so let's stick to two or three dimensions for the time being. Take another orange and a black marker. Now choose a natural number, let's say n, and try to draw n dots on the orange in such a way that the minimal distance between these dots is maximised. If it weren't for this last requirement ('some distance needs to be maximised'), this task would be utterly trivial: just put all the dots very close to each other. But the minimal distance between two arbitrary dots should be maximal. Think of a teacher's philosophy during the exams (maximising the distance between his pupils) versus how the kids themselves would organise it if they could (all sitting next to each other).

If the problem with the orange is too hard, why not start with spherical codes in two dimensions first? Spheres then become circles, which means that spherical codes in two dimensions can be seen as table configurations at wedding parties. Well, this suddenly sounds like we're dealing with an even harder problem now — I have the impression that some wedding seating plan arrangements require more planning than a six-month solo survival trip to Antarctica — but trust me, it's really not that hard. If you are to seat n people around a (circular) table in such a way that the extra requirement is satisfied, the only solution is to place the chairs at equal angles (you can even say which one: 360 degrees divided by n). For any other configuration, there will be at least two people sitting closer to each other than all the others, which means that the minimal distance is not maximal (you had *one* job).

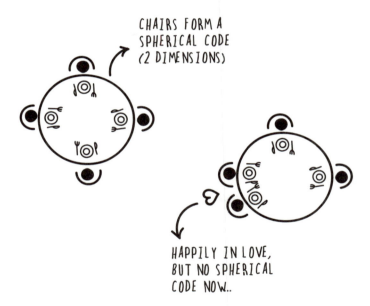

CHAIRS FORM A
SPHERICAL CODE
(2 DIMENSIONS)

HAPPILY IN LOVE,
BUT NO SPHERICAL
CODE NOW..

Wedding table engineering

Now back to our orange in three dimensions. Once again, the goal is to draw n dots on the orange in such a way that these are all as far from each other as possible. For $n = 2$, this is easy: draw the first dot wherever you want, and put the other dot on the opposite side. The case $n = 3$ is also not that hard, especially if you keep in mind that three points always fix a plane (a mathematical fact which, sadly enough, has not yet found its way to the people designing pub terrace tables[66]). For $n = 4$, it suffices to put one person at each vertex of a regular pyramid inside the sphere.

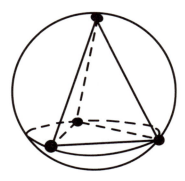

And for those of you who are wondering what mathematics has to say about your personal bubble space: you might want to calculate the spherical code in three dimensions for n roughly equal to 7.7 billion (the current world population). You do not even need to do it for real I guess: tapping someone's shoulder and saying, 'Excuse me, I've determined the size of the three-dimensional spherical code of order 7.7 billion. Would you like to hear it in radians, or shall I convert it into degrees for you?' usually buys you some extra space (and a disturbed face from the owner, desperately trying to work out whether you're just friendly, or condescending).

---

66 Seriously, with all the beer coasters I have been folding to balance tables in the past, I can probably plaster a building.

Recommended listening

Artist	Song title
Amenra	Plus près de toi
American Nightmare	I've Shared Your Lips So Now They Sicken Me
Ben Frost	The Teeth Behind The Kisses
Slipknot	Wait and Bleed
Innerwoud	Nachtkus
Stabbing Westward	Waking Up Beside You
Red Hot Chili Peppers	Suck my Kiss
How to Destroy Angels	The Space in Between
Pet Shop Boys	Integral
Integrity	Nothing Left

5

Space To Choose

> *Choose Life. Choose a job. Choose a career.*
> *Choose a family. Choose a fucking big television,*
> *choose washing machines, cars, compact disc players*
> *and electrical tin openers.*

(Irvine Welsh)

You should really read this chapter if ...

- your secretary can be a bit of a problem too sometimes.

- the story about Schrödinger's cat doesn't make sense to you.

- you haven't seen your friend lately (the one who has two children).

- you really want to beat your partner in that weekly Game of Googol.

5.1 Is choosing really losing?

One place where gently kissing the boundaries of one's personal bubble can typically turn into sheer space invasion is a music festival. Having once spent a month in India, I realised that there are actually lots of similarities between attending a music festival and travelling in the second-most populous country in the world.

Definition 31.

Music festival: *an organised series of performances, typically held in the same place on an annual basis, characterised by the fact that there will always be at least one person present who is taller than you. No matter where you are on the festival ground, this person will find the spot right in front of you and block your line of sight for the rest of the concert. The Law of Conservation of Misery applies here: moving to the first row never works, as there is also at least one security guard or journalist present who is taller than you. If this is not the case, an oversized stage monitor will do the trick.*

Theorem 2. *Music festivals are miniature copies of India.*

Proof: First of all, entering either of these places can be quite a hassle: for most music festivals you need to to buy an entrance ticket, and in order to enter India you must apply for a tourist visa. Both tend to be overpriced nowadays and unless you really want to push your luck you'd better not buy them on the street from a guy showing you the inside of his jacket as if it were a display case. Moreover, as anyone who has ever spent a few hours on the toilet with a book, a new toilet roll and a serious bowel obstruction can confirm: bringing the appropriate pieces of paper is by no means a guarantee of success.

At the gates of the festival camping, your luggage is usually screened by a team of sturdy security guards holding at least a bachelor's degree from the MacGyver Institute of Technology: anything that even remotely looks as if it could serve to help you build an improvised cross between a hovercraft and a nuclear warhead launcher is removed and thrown into a container, together with your drugs, knives, BBQ-sets and bottles of alcohol — waste collectors also have extralegal advantages, I suppose.

At the border crossing into India, it is not just your luggage that will be scrutinised: you also need to get through immigration yourself, a procedure which involves being intensely stared at by an officer switching from your face to your passport as if she is playing a game of 'find the seven differences'. Next time I need passport pictures I think I'll go to a railway station photo booth having spent the night on a slab of concrete, in a room in which the air is so dry that it would make a bag of powdered bones look moist — just to make sure I look like myself when I am facing the airport immigration officer.

Once you have made it inside, it feels like you have just set foot on another planet. A crowded one that is, because the first thing you will notice is way too many human beings. People in front of you, folks behind you, blokes to your left and to your right. Depending on your opening lines and preferences, they may even end up underneath or on top of you. I wouldn't be surprised to hear that there are universities in India where you can attend a course entitled 'Packing People into Public Places', for they surely seem to have taken this concept to a completely different level: I remember once taking a local bus from Mahabalipuram to Pondicherry which was so crowded that I was no longer able to tell where my own body ended and my neighbour's began. Distraction was provided, though, in the shape of a colourful Bollywood flick on the shabby television screen at the front of the bus. This didn't really work for me though; I had read in my guidebook that Indian movies are full of sexual references, and the last thing I needed on that bus was collective arousal.

On top of that, there seems to be a cosmic principle which says that whenever you put too many people in one place, there will be a mechanism at work to ensure that you end up with your nose in a sweaty armpit. On buses this is accomplished by installing grab handles, whereas at festivals you can always count on the combination of your neighbour's enthusiasm and his evident lack of feeling for rhythm. Whenever there is a beat involved, he will raise his arms over his head to begin his irritating syncopation. When there is no beat, he will do the same and start taking pictures with his phone.

Either way, you will be exposed to his rank oxters. Where is that *deo ex machina* when you need it?[67]

As for my final argument, I would like to point out that both the Indian subcontinent and music festivals are places inhabited by people nodding their way through the day. At rock music festivals this is either because people are headbanging, or because party drugs took control and turned the owner of the head into a life-sized bobbing head dashboard doll. In a car with bad suspension. Driving on a cobblestone road. In a mountainous region.

In India, people tend to make a distinct head movement because they just can't say 'no'. In case you think I am merely exaggerating personal observations, there is an actual Wiki page dedicated to the head bobble. This obviously leads to social situations which go from slightly confusing to downright upsetting. When I was travelling in India, I tried to demystify their tottering, asking people on the street: 'If you wobble your head, does that mean no?' They just wobbled their head.   Q.E.D.

Definition 32.

Indian head shake: *the true power of the head bobble can only be properly understood using the language of quantum mechanics. This is an advanced theory in physics which focuses on the behaviour of matter and energy on the elementary level of subatomic particles. One of the best-known experiments in quantum theory is known as 'Schrödinger's cat'. It involves a cat in a box — nothing special here, it is a cat after all — which is stuck in a state that I myself can only achieve after 37 bottles of vodka in a room filled with toddlers testing vuvuzelas: dead and alive, at the same time. Physicists then explain this quirky behaviour by saying that 'the cat sits in a quantum superposition of two states'. I think part of the reason why many people don't understand quantum mechanics is because this is a bad example, which goes against our intuition. The thing is that*

---

67   Indeed, thrown into a container by the diligent security guard.

quantum superpositions do exist, but there is absolutely no need to go to the subatomic level: just ask an Indian person a yes and no question, and you'll get your quantum behaviour.

There is actually one more thing that rock festivals and India have in common: unless you are a member of the upper castes (who would have thought that 'VIP member' is an ancient Indian concept?), going to the little inventor's room can be quite the adventure. Either because Delhi is laying claim on your belly *and* bowels, stressing its self-proclaimed sovereignty by the forced complete evacuation of the native population, or simply because there's a toilet cubicle queue which is so long that by the time it is your turn you could easily forget why you were waiting in the first place. If it weren't for the pressing answer in your pants, that is.

Despite it sometimes being a suspiciously small space — when I lived in Tokyo, the toilet in our apartment was so cramped that I expected to go back home as a certified contortionist, if not chronically constipated — it is the sort of space we prefer to be of the spotless kind. So if you are like me, and the first thing you do upon entering the festival grounds is to make a run for the johns, beating the crowds who are crawling out of their tents like prisoners coming out of solitary confinement — wondering how and when they ended up in there, not yet accustomed to the broad daylight — it can be very useful to know how you can optimise your chances of finding a plastic portable toilet which does not smell like a colony of incontinent wombats just died in it.

Not surprisingly, there does indeed exist a mathematically sound answer to the question: how to maximise the chances of finding the best candidate amongst a certain number? A strategy which can be used not only when you are facing a row of toilet cubicles; it can also be used for all sorts of 'selection problems', such as buying a house or hiring people (whence comes the name under which it appears in mathematical textbooks: *the secretary problem*).

Definition 33.

The secretary problem: *most mathematical textbooks refer to the problem described above as the secretary problem, but this is a common misconception. The secretary problem actually refers to the following long-standing problem in optimisation theory: 'Given that your inbox receives 314 emails per day requiring immediate action (15% of which contains animated GIFs displaying a kitten wearing a party hat that need to be shared with at least 92 other people on Facebook or Twitter), given that 65% of your colleagues will casually drop by your office to engage in a gossip exchange that will take 3.58 minutes on average, given that you have to brew 19 litres of coffee for 11 meetings in four different offices which are separated by 2.9 kilometres and three cleaners complimenting you on your shoes, and given the fact that you will have to leave work 42 minutes early to pick up your five-year-old son from school, calculate the speed with which you can finish a low-carb vegan 320-calorie lunch.'*

In order to tackle this problem of maximising our chances to make the best choice, we will make the following assumptions:

(i)   The total number of candidates from which you have to make a choice is known in advance. Let us call this number N. Going back to the situation sketched earlier (trying to find the cleanest festival toilet), this means that you will have to count the number of cabins. Trust me, when it comes to finding a spotless one, it still beats counting *on* them.

(ii) There is a unique best candidate amongst the lot and this is the one we would like to single out, using a decent strategy. The winner takes it all, it's simple and it's plain.

(iii) You can see the candidates one after the other, in random order, but you have to make an immediate decision once you think you've seen the best candidate. If you were to omit this rule, it would of course be a much simpler ball game: you just try them all, and then pick the best one at the end of the selection.

One possible strategy is of course to make a purely random choice. There is a chance that you accidentally choose the best candidate, but this chance is rather slim. Let's say that $P_r(N)$ stands for 'the probability of randomly picking the best candidate out of N possibilities'. As there are N candidates in total — one of which is better than the rest — the probability of randomly picking this one out of the pack is given by the number

$$P_r(N) = \frac{\text{unique best candidate}}{\text{total number of candidates}} = \frac{1}{N}$$

For instance, picking a random candidate from 20 people during the job interview gives you a 5% chance that this is the most suitable candidate. We can actually do much better than that, adopting the following strategy:

- We will look at a certain number of candidates (let's call this number t for 'trial') and remember the best candidate amongst these. In a sense, this is our sample group, used for future comparison. Note that we will not choose any of these candidates — no matter how impressed we are — we merely use them to set our benchmark. I have once used this as an excuse to dump a girlfriend: 'Sorry, darling,' I told her, 'I am applying the secretary problem's strategy in my search for the perfect partner.' 'That's okay,' she answered defiantly, 'you were not bold enough to be my type anyway. But at least it explains why you were always bossing me around.'

- After that, from candidate $(t + 1)$ onwards, we will simply choose the first candidate which is better than the best candidate from the t we have just seen. This is really the crux of the idea: because we do not know which of the N candidates is the overall best, we first create ourselves a sample group of t people, and the best of these will then serve as the one to beat.

After which everything reduces to crossing the fingers and hoping that, amongst the rest of the candidates, there is someone who's indeed better than our benchmark.

- If no such candidate is found (which then basically means that the benchmark accidentally happened to be the best of *all* the candidates), we obviously have to choose the last one.

Just like that guy who never saw boiling water being poured over cured leaves, you may ask yourself: 'What is this thing you call t?' Because I hope you can all agree that the size of this sample group will play a crucial role. So far, we have not specified its value; it is just a number less than N, the total number of candidates. It is a very good question indeed, but I will postpone the answer until later in this chapter. As for now, you can think of it as a random number between 1 and N. Let us first calculate the chance of finding the best candidate, given this refined strategy. It is given by the following formula:

$$P_s(N, t) = \sum_{j=1}^{N} P(\text{candidate } j \text{ is the best and is chosen})$$

Yet another formula, so before your maths phobia grows completely out of control and has you running around with a terrible rash, both eyeballs turned inward, barfing green bile containing chunks of honey glazed ham, grunting Justin Bieber songs backwards in an extinct Ukrainian dialect, defiling the wall-paper with satanic messages written in the blood of decapitated hamsters, allow me to explain what this piece of boxed sorcery stands for.

Remember that we are looking for a total probability here, so I just gave it a name: $P_s(N, t)$, where 's' stands for 'strategy' — or, depending on your mood, 'seriously, more of these stupid symbols?'. It is given by a sum, as can be seen from the notation: the capital Σ, with subscript j going from 1 to N, tells us that we are adding N numbers in total — which all somehow depend on this index j, which basically labels the candidates. Put differently: there is one term for each candidate. As a matter of fact, each of the terms in the sum at the right-hand side of the equation should be read as the probability that candidate j is the best and is moreover chosen. Well, it can also be read as an abstract haiku, but only if you have a really crooked sense of poetry.

Definition 34.

Haiku: *form of poetry, looks like Master Yoda quote, often less funny.*

In order to see why we need to *add* the probabilities here, think of the following example: the probability that you were born on a Monday *or* on a Friday is equal to a sum of probabilities: $\frac{1}{7} + \frac{1}{7} = \frac{2}{7}$. Unless you are Jewish, because there's no labour on Saturday, so then the answer would be $\frac{2}{6}$ or $\frac{1}{3}$.

How do we now calculate this probability for candidate j, to be the best and to be chosen? If there is one thing that I have learnt from televised talent shows in which the audience determines the winner through text messages, it is that 'being the best' does not necessarily imply 'being chosen', so we will have to take both probabilities into account here. To be more precise, we need a formula for calculating the probability that two events A and B will happen simultaneously. That, and hugs — wel all need more hugs.

First of all, if the events A and B have nothing to do with each other, then the probability that both things happen is the product of their respective probabilities. This can be expressed as

$$P(A \text{ and } B) = P(A) \times P(B).$$

One may wonder why we need to multiply these probabilities as opposed to adding them — or having a unicorn serve them rainbow soup with meatballs; the clue to solving a mathematical problem might be to treat it well — so let me give you an example: if there is a 50% chance that it will rain tomorrow, and a 50% chance that the book you ordered through Amazon will finally arrive, then there is a 25% chance that your postman will deliver a soggy package tomorrow, as $\frac{1}{2} \times \frac{1}{2} = \frac{1}{4}$.

However, the situation becomes more complicated when the events are 'connected' somehow. In that case, one should use *conditional probability*, a statistical theory centred around the idea that the occurrence of an event *A* sometimes encodes information which restricts the chances that event *B* will also happen. If Bayes' name shot through your brain upon reading the previous sentence, someone conditioned it well: his theorem has everything to do with it.

Definition 35.

Conditional probability: *this is one of the most fundamental concepts in probability theory. It should not be confused with* conditioner probability, *a theory trying to work out why a doorbell is more likely to ring when the hair of the owner is all soaped up under the shower.*

For instance: suppose recent studies have shown that (A) the probability that a man wears white socks on a first date is 10% (B) the probability that a man has sex on a first date is 20%.What is then the probability that a man wears white socks on his first date *and* has sex? According to our naive formula from above, that should be

$$P(A \text{ and } B) = P(A) \times P(B) = \frac{1}{10} \times \frac{1}{5} = \frac{1}{50} ,$$

which is 2%. And yet, we all know that this is plain wrong, as women don't find that very attractive. The answer should obviously be equal to zero. Does this mean that our formula from above is wrong? Of course not. It is still correct, but it should be used in the right situations only. If you comb your hair with a straight razor, you shouldn't question the device, but your strategic judgement skills.

For those situations in which the outcome of event A (wearing white socks) may influence the outcome of event B (having sex), one has to use a different formula. It is the cornerstone of conditional probability, and it says that $P(A \text{ and } B) = P(A) \times P(B|A)$, where the latter should be read as 'the probability that the event B will happen if you know that event A has happened'. The nice thing is that this formula actually *refines* the previous one: if A and B are independent, then $P(B|A)$ is simply equal to $P(B)$ because knowledge of whether or not A has happened does not change anything in that case.

Going back to our example, we simply have that $P(B|A)$ — the probability that a man has sex on a first date, given the fact that he is wearing white socks — is equal to zero. As a side note, there is ample space to choose from if you do want a carnal topping on the first date soup: any other colour of socks will do. Unless you are dating someone who adheres to the philosophy that the set of white socks contains white socks *and* sports socks — regardless of their colour. An ex-girlfriend of mine fell under that dubious category. She broke up with me when I started wearing white sports socks; little did I know that negative times negative is not always positive.

In many textbooks, conditional probability is often explained in terms of people having children. Suppose you bump into a friend you have not seen in a long time. Assuming that boys and girls are equally likely to be born — which is not true, demographically speaking, but this is a mathematical idealisation — what is the chance that she has 2 boys? This case is covered by the classical rules of probability: as the events 'child 1 is a boy' and 'child 2 is a boy' are independent of each other, we can use our first rule to calculate the total probability. Event A (the firstborn is a boy) occurs with a 50% chance, and so does event B (the second child is a boy). This means that

$$P(\text{child 1 is a boy and child 2 is a boy}) = \frac{1}{2} \times \frac{1}{2} = \frac{1}{4}.$$

Now suppose you meet a friend you haven't seen in a long time, holding a boy's hand.[68] If you know that this is one of your friend's two children, what is the probability that she has two sons? A common mistake is to think that this is still 25%, again assuming that boys and girls are equally likely. However, the fact that you can see that he has at least one son changes everything. Simply put: two daughters is no longer an option, which means that there are only three possibilities left (boy-girl, girl-boy and boy-boy). So the correct answer is a little over 33%.

---

68   Well, and the rest of the boy's body attached to it of course — you psycho.

The three remaining options

One more example? Suppose you meet a friend you haven't seen in a long time — just in case you're wondering why this is always part of the problem, don't forget that they have children — shouting 'Put down that hamster, we are not going to bite off its head this time!' at a boy with an eye-patch and a blood stained t-shirt. If you know that this little creature is your friend's son, what is the probability that she has two boys? The answer again follows from the rules of conditional probability: $P(2 \text{ boys}) = P(A) \times P(B|A)$, where $P(A)$, the probability of having a boy, is equal to 50% and $P(B|A)$, the probability of having a second boy when you gave birth to an ambassador of Satan, is of course equal to zero. Not making that mistake twice, right?

Now that we have acquainted ourselves with the basic rules of conditional probability, it is time to return to our secretary problem. You should now agree with me that the final probability $P(A$ and $B)$ that candidate j is the best (event A) *and* will be chosen (event B), is equal to the probability $P(A)$ that he or she is indeed the best, multiplied with the *conditional* probability $P(B|A)$ that this person is picked *given* that he or she is indeed the best. So this means that the number $P_s(N, t)$ we are after is equal to

$$P_s(N, t) = \sum_{j=1}^{N} P(j \text{ is best}) \times P(j \text{ is chosen} \mid j \text{ is best}) .$$

It's the same horrible sum as above, labelled with the index j ranging over the total number of candidates, but the probabilities at the right-hand side of the equation were rewritten. My professor in Theoretical Mechanics used to say: 'Mathematicians don't solve problems, they just rewrite them until they look nicer.' This is exactly what we have done here: we still don't know how likely it is that our refined strategy leads us to the best possible candidate. However, we did manage to break up our problem into feasible chunks. Think of the stategy you adopt in the kitchen: first you make the sauce, then you boil the pasta. Put them together, and you are one step closer to solving your original problem: getting rid of that huge pile of half-opened packs of spaghetti in your pantry.

The first chunk is the easiest one: as all the candidates are equally likely to excel, we still have that

$$P(\text{candidate } j \text{ is best}) = \frac{\text{unique best candidate}}{\text{total number of candidates}} = \frac{1}{N}$$

The probability that we will also *choose* this candidate is a bit more difficult to explain, as it depends on our strategy — fixed by choosing that number t, the size of our sample group from which we select our benchmark. For the first t candidates (labelled by an index $1 \leq j \leq t$), we simply have that the probability that this candidate is chosen is equal to zero. Nothing personal, we just decided to not choose him or her as part of our initial strategy. As for any of the other candidates, labelled by an index between $(t + 1)$ and N, there is only one thing that can go wrong — well, two if you include being hit by a low flying drunk piano, but we usually ignore this factor. It could be that candidate j is indeed the unique best one, but that there is a candidate coming before j who is already better than our benchmark. In that case we would obviously pick the wrong candidate, as the strategy dictates that we choose the *first* candidate who is better. For example, take a look at the situation below: I have lined up ten candidates ($N = 10$), and expressed their rank as a numerical score (with 10 being the best result). Let's assume that our sample group consists of four people ($t = 4$):

Candidate	1	2	3	4	5	6	7	8	9	10
Score	2	4	7	1	6	5	8	3	**10**	9

The best candidate amongst the first four is clearly candidate 3, with a score of 7. So what our strategy now dictates is that we should go for the first candidate who beats this score, but unfortunately enough this means that we will wrongly go for candidate 7 (with a score of 8) instead of the overall best candidate, at position number 9.

This example shows that our strategy does not always work, so what are the odds that it *does* work? To answer that question, we must calculate the odds that candidate j is indeed picked out, given that he or she is the best one. The rest of this sentence may require rereading,[69] but it all amounts to this: the *second* best candidate amongst the first j candidates should be amongst the first t. Have you really read it twice? Just once more for good measure, okay? *The second best candidate, when counting up to j, should be amongst the first t.* In our example above, the best candidate has a score of 10. The highest score before this candidate is 8 (not 9, mind you, as this comes *after* the best candidate), so our strategy would work perfectly if this score would be amongst the first four candidates. This remains true in general: we first count up to t, retaining the best candidate of our sample group. If this person happens to be the second best amongst the first j candidates, then we can safely continue until we reach candidate j. As this is *the* best candidate, it is obviously the first (and only) one we will meet that is better than the benchmark.

So, the final piece of our puzzle: what is the probability that this *second best* candidate amongst j has a label between 1 and t? Think of it this way: you have two prizes to award, and j candidates. The last one (candidate j) gets first prize. Second prize should be awarded to any candidate of your choice, as long as he or she has a label between 1 and t (it must be someone in the sample group). There are $(j - 1)$ candidates left (you already awarded the first prize to the last one), but only t of these can be chosen. All in all, this means that the probability P(candidate j is chosen j is best) that we were after is equal to

$$\frac{\text{the people in the sample group}}{\text{all the people before candidate } j} = \frac{t}{j-1}$$

---

69   I find 'reread' a funny word: it is spelled in such a way that merely reading it feels like you are already doing it.

Definition 36.

Puzzle: *the most popular example is the jigsaw puzzle, a tiling puzzle which consists of oddly shaped interlocking pieces of paperboard or wood. The final part of a problem is also often referred to as 'the last piece of the puzzle'. Given the fact that it is pretty obvious where the last piece of a puzzle goes — just like gay penetration sex: it's not like you have an option — I have always wondered why they don't call it 'the penultimate piece of the puzzle'. At least this one still requires some work.*[70]

Third time's a charm, so let us for the last time return to the problem we were trying to solve (probabilities and patience, both spelled with a capital P). We've calculated the conditional probabilities involved, so we can plug these into the formula for the total probability $P_s(N, t)$ that we will pick the best candidate amongst N with our sample size strategy (the beast in the box from a few pages ago):

$$P_s(N,t) = \frac{1}{N} \sum_{j=t+1}^{N} \frac{t}{j-1} = \frac{t}{N} \left(\frac{1}{t} + \frac{1}{t+1} + \ldots + \frac{1}{N-1} \right).$$

Still a sum, but as the conditional probabilities have been plugged in, we finally obtained a formula that we can explicitly calculate. The number $P_s(N, t)$ obviously still depends on the number N, the total number of people which was known right from the start, and the number t, which we haven't specified yet. 'But how can I calculate something if I don't know what t is?' Well, this is where we can start experimenting: t can be any number between 1 and N, so we can calculate the number $P_s(N, t)$, using the formula above, for all possible values of t.

---

70 If maths itself has any sense of humour, the hardest piece of the puzzle is the one you have to put in when 37% is finished.

The upshot is that as we are trying to *maximise* the probability of picking the best candidate, the smartest choice for *t* would be the one for which this value $P_s(N, t)$ becomes maximal.

Let us consider our example first: suppose there are $N = 10$ people. If we just pick someone randomly, we have a 10% chance that this person is the best one. But what happens if we use the formula above? In other words, we first choose a 'benchmark size' (the number *t*) and then calculate the probability $P_s(N, t)$. For instance, using the formula from above for $t = 3$ (substituting the letters with numbers), we arrive at the following probability:

$$P_s(10, 3) = \frac{3}{10} \left(\frac{1}{3} + \frac{1}{4} + \frac{1}{5} + \frac{1}{6} + \frac{1}{7} + \frac{1}{8} + \frac{1}{9} \right) \approx 39.87\% .$$

Listing these probabilities for *all* possible values for *t* (doing the sum ten times in total) gives the following approximate percentages:

t	0	1	2	3	4	5	6	7	8	9
$P_s(10,t)$	10%	28%	37%	40%	40%	37%	33%	27%	19%	10%

First of all, we can see that for the extremal values for *t* we again get a 10% chance. This is easy to explain: for $t = 0$ there is no benchmark to compare with, so we simply pick the first candidate available.[71] For $t = 9$, we are left with picking the last one as there is only one candidate left. But as you can see from the numbers, we can definitely do better than 10%: with 40% for *t* equal to 3 or 4 we are getting pretty close to flipping a coin (a 50% success rate).

---

71   This somewhat resembles the philosophy some people seem to adopt on a Friday night in a club, when bad sex is preferred to no sex.

One can even prove, although this requires more advanced machinery, that the optimal benchmark size (the number t in our strategy) is given by

$$t_{optimal} = \frac{N}{e} \approx \frac{N}{2,7182818284590...} \approx N \times 0.37$$

The number e in the formula above is Euler's number, which we have already met in an earlier chapter. Like a crumpled banknote in a pocket of jeans you haven't worn in a long time: it is completely unexpected, but in a good way. Because for some bizarre reason, numbers like π and e have the same effect on a mathematician as an offshore bank account on a tax inspector: a forewarning that they are onto something good. So roughly speaking, the best strategy is to ignore 37% of the candidates (when $N = 10$, as above, this gives $t = 4$ as you can see from the table) and to compare the remaining 63% with the best of those. Moreover, by adopting this strategy (a sample group with 37% of the total amount to compare with) the probability of effectively choosing the best candidate will be given by

$$P_{optimal} = \frac{1}{e} \approx 37\%$$

This may not seem a lot, but it should be contrasted with the random choice, which gets smaller and smaller as the number of candidates increases.[72] Our optimised strategy roughly works a little more than once every three times — regardless of the number of candidates. If only 'ordering the last drink of the evening' had the same success rate, that would have saved me a few nasty hangovers.

---

72   As far as I know, the only other example of a probability that doesn't become smaller when the size of the group increases, is the chance that there will be at least one person who doesn't know that 'scampi' is actually already a plural.

5.2 Brainy with a chance of pitfalls

5.2.1 Seasoning adds complexity

I honestly believe that probability theory and statistics are the psychedelic drugs of mathematics: constantly messing up our minds. I mean, a statement like 'on average, people have less than two legs' immediately has our brain in a state of bewildered excitement because it sounds wrong, but it's a simple fact of life. One of the best examples I know of is a more than fascinating consequence of the theory of conditional probability which is known in the literature as the 'Boy or Girl Paradox'. The mathematical conundrum, that is, not Conchita Wurst. It also goes under the moniker of 'the Two Child Problem'; I bet the Chinese are glad to read that they are not the only ones struggling with that. At this point it probably does not surprise you anymore that the aforementioned paradox again involves the offspring of a mate. One that you, once again, haven't seen in a long time — at this point I am starting to believe that the version of you which appears in probabilistic puzzles has some sort of social issue.

Imagine him holding hands with a young girl (it's a boy in the standard textbooks, but we do need more women in maths so we might as well start here). He also tells you that she is one of the two children he has, and that she was born in the autumn. We hereby tacitly assume that the chances of being born in any season are all equal to 25%, although this may again not be true demographically speaking: in Ireland, for instance, fewer people are born during the summer (then again, it is still the best day of the year). Anyway, given these facts, what is the probability that your friend has two daughters?

Definition 37.

Birthday: *an annual celebration of one of the most — if not the most — embarrassing day of your life: you are being pushed out of a vagina, bloody and naked, and ass-spanked in front of an audience until you start crying. Just in case you are wondering: yes, I was born in a hospital, not in a Berlin nightclub.*

Apart from the gender, we had this question before, so it's only normal to think that the answer is again $\frac{1}{3}$. However, the problem considered here is slightly different: this time we also know that the daughter you met was born in the autumn.

'How can that be relevant information?'

That's exactly what I thought when I first encountered this question, so I understand your confusion. Please bear with me. There are three ways to approach this problem: we can use Bayes' formula — the milestone in the history of maths we mentioned in the previous section — we can undertake a five-year-long voyage to a mystical mountain in China, where a long-bearded monk with a shiny head who lives on a diet of broken dreams and rusty nails will wrap the answer in a Buddhist koan, or we can simply count. Because this is what calculating a probability boils down to: out of *all* the possible outcomes, you just count the ones that correspond to the result you are interested in. If I keep a natural number between 1 and 50 in mind and you can have one guess, there is a 2% chance that you will guess correctly: 50 possible outcomes, one of which is correct.[73]

---

73   Connoisseurs have already worked out that it's *way* more than that of course: there's a 99% chance I will be thinking about the number 42.

	Spring B/G	Summer B/G	Autumn B/G	Winter B/G
Spring b/g	Bb Gb Bg Gg	Bb Gb Bg Gg	Bb **Gb** Bg **Gg**	Bb Gb Bg Gg
Summer b/g	Bb Gb Bg Gg	Bb Gb Bg Gg	Bb **Gb** Bg **Gg**	Bb Gb Bg Gg
Autumn b/g	Bb Gb **Bg Gg**	Bb Gb **Bg Gg**	Bb **Gb** **Bg Gg**	Bb Gb **Bg Gg**
Winter b/g	Bb Gb Bg Gg	Bb Gb Bg Gg	Bb **Gb** Bg **Gg**	Bb Gb Bg Gg

So let's try a counting argument for our two-daughter problem: out of *all* the possible outcomes, we only need to single out the ones we're interested in, and then do the division. To make this somewhat easier to follow, take a look at the table above: it is a graphical representation of all the possible situations. The letters B/G (boy/girl) are used for the oldest child, and b/g for the youngest. Any square, considered as the intersection of a row and a column, contains four entries: an older boy or girl, born in the vertical season, and a younger boy or girl, who saw first light in the horizontal season. These are the standard four possible combinations of two children. Now, as we know that there is at least one daughter born in the autumn, we can focus our attention on the third row (this would mean the youngest was born in the autumn) and the third column (the oldest was born in the autumn). The other squares can be ignored. Moreover, we can also ignore the entries 'Bb' in the remaining squares, since there is at least one daughter. In other words, it suffices to focus on the bold-faced entries. This is the point where we can start counting: there are exactly 15 cases in which at least one girl was born in the autumn (the third column for the older one, plus the third row for the younger one), and only in seven of these cases are both children girls (**Gg**). All in all, this comes down to a probability of $\frac{7}{15}$, a little less than 50%.

If this counterintuitive result just threw you off balance, you can take comfort in knowing that you are not alone: mathematicians, statisticians and other professional thinkers are actually still struggling with the outcome of this birthday conundrum. The numbers obviously don't lie (unless you happen to be the president of Brazil and the numbers come from the head of the agency tracking Amazon deforestation), but the idea that seemingly irrelevant information can have such a striking impact on the probabilities is a baffling one.

The original version of this probabilistic puzzle was presented by Gary Foshee (a recreational mathematician and puzzle designer) at the Ninth Gathering 4 Gardner Conference in 2010 (I adapted it slightly to make the analysis a little easier to follow). At the time, this problem created a huge wave of astonishment in the participating audience, and since then it has spawned intensive discussions on the internet, with plenty of highly educated people dedicating articles to it — most of which are like avian travel routes: going way over my head. I suppose once you have devised a mathematical puzzle in which both cognitive psychologists and philosophers have taken an interest — because they want to study it from the point of view of human perception, or because they start questioning the definition for what 'a probability' really is — you can safely tick 'keep the academic world busy for a while' off your bucket list.

Definition 38.

Birthday Paradox: *experts often think that this term stands for a variety of probabilistic puzzles involving sons and daughters. In reality, though, there is only one true Birthday Paradox: if you give a toddler a wrapped roll of wrapping paper as a birthday present, what will it choose to play with?*

5.2.2   Even odds can be odd

When it comes to probabilities messing with our intuition, the 'Game of Googol' is one of my absolute favourites. For the people who have skipped the chapter 'Setting Up Space': Googol is a moderately large number. For people suffering from short-term memory loss: Googol is a moderately large number. For all the people who overestimate the power of the comic triple: too much indeed. As the American mathematician Thomas Ferguson argues in his paper 'Who solved the secretary problem?', the Game of Googol problem is — historically speaking — the true secretary problem, although the original formulation of this conundrum has as much to do with secretaries as the 'happy ending theorem' with massage parlours.[74]

It first appeared in print in 1960, in Martin Gardner's column in *Scientific American*, and it goes as follows: 'Ask someone to take as many slips of paper as he pleases, and on each slip write a different positive number. The numbers may range from small fractions of 1 to a number the size of a Googol (hence the name) or even larger. These slips are turned face down and shuffled over the top of a table. One at a time you may start turning the slips face up, and the aim is to stop turning when you come to the number that you guess to be the largest of the series. You cannot go back and pick a previously turned slip. If you turn over all the slips, then of course you must pick the last one turned.' The similarities with the problem treated in this chapter are obvious, which means that the player turning the pieces of paper has a strategy to identify the biggest number which will work with a probablility of 37%.

---

74   This theorem says that for any five points in general position in the plane (so maximum two on the same line), four of these will always form the vertices of a convex quadrilateral. Most people tend to be happy when they reach the end of the description of this problem, which may explain the name.

However, Ferguson points out what he believes to be a crucial difference between Gardner's version of the secretary problem (the Game of Googol) and the one treated earlier in this chapter: the latter only talks about a *relative* rank separating the candidates (they don't come with numbers, they can only be compared to one another). In the former, we are dealing with *absolute* values, and there is another player involved who can actually *choose* these numbers. Just like all the psychological and philosophical arguments which appeared in the aftermath of Foshee's seasoned birthday problem from the previous pages, this is about as subtle as it can get. Then again, sometimes it's all about the details — a lesson that some people learn in certain bars in Thailand, when the ladyboys' specifics are overlooked (another happy ending problem[75]).

What we are really dealing with in Gardner's formulation is a two-person game and this means that it should be attacked using game theory techniques. Now instead of solving the Game of Googol in full generality, for an arbitrary number of paper slips, we will take a look at the simplest possible version of this game, involving two numbers only. So, player A jots down two completely arbitrary numbers, as big as he wants, and player B can turn over one piece of paper. As his aim is to pick the biggest number, this means that he is facing the following problem: should I stay, or should I go? Should you go for the number on the piece of paper you've just turned over, or take a gamble and choose for the other piece of paper? There is a 50% chance that the invisible number is bigger, so this sounds like a flip of a coin — right? Well, not exactly: what if I told you that there is a strategy which gives you *better* odds than that?

'Burn that man! He is a witch!'

---

75   I once went to a massage parlour in Vietnam. At some point the masseuse started doing my lower body, and said 'Sir, you are so big down here!' I sheepishly thanked her — how do you react when someone has unintentionally sexually aroused you — but she started smiling and added 'Haha, I was just pulling your leg, sir.'

And yet, it can be done. I found this beautiful solution, based on an idea from Greg Egan. However, you will need a few things for this: a piece of paper, a pen, a pair of scissors, duct tape, a ruler, cardboard and a baseball pitching machine (or any other cannon which fires balls[76]) which will serve as a random number generator. First of all, take pen and paper, and draw the graph of a strictly increasing function, like the one in the picture below.

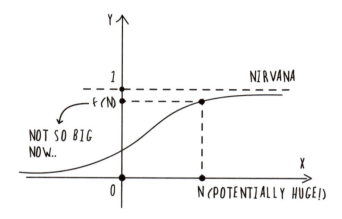

In sharp contrast to most professional athletes' abs, the actual shape of this function is not too important. The only thing that really matters for our purpose is that it satisfies the following properties, which can be observed from the graph:

- For any real number x on the horizontal axis, the corresponding value $f(x)$ must be a number somewhere between zero and 1 on the vertical axis. So the good news is that your piece of paper may have a finite height of unit length.

---

76   If you don't know how to make this, you can always hire the A-Team and lock them in a garage with a welding machine, two oil drums, a few PVC pipes and 12 boxes of canned peaches.

- The bad news is that your piece of paper will have to be infinitely long, because *all* real numbers x (living on the horizontal axis) must have a corresponding value $f(x)$. Moreover, you cannot draw your curve too wildly, you have to respect some kind of Zen principle ('aiming at, but never reaching'). What I mean by that is the following: at the left-hand side of the vertical axis, the graph curves towards the horizontal axis. So, the more negative x becomes, the closer $f(x)$ will approach zero (from above). On the other side of the vertical axis, number 1 is Nirvana: the bigger x becomes, the closer $f(x)$ will get to the number 1. Without *ever* reaching this value, a Buddhist principle in mathematics which can be expressed in terms of limits. Judging from the number of people who are still being treated to a speeding ticket, this concept might be a bit too difficult though.

- Finally, your graph is not allowed to have peaks and valleys, it should be *strictly increasing*. This means that if x increases, the value $f(x)$ increases too. Like climbing the stairs of a holy mountain in Asia: you can do it at your own pace — the steepness of the curve is completely up to you — but you can't go back.

The main idea behind our funky strategy is the following: start by randomly choosing a piece of paper — no surprise there of course — and flip it. You are now staring at one of the numbers your opponent scribbled down, let us call it n for the time being. This can be a huge number — even one of those giant beasts for which mathematicians developed their own notation because there's simply not enough space in the universe to jot them down — but don't let this deter you: you can easily tame this beast, using the graph that you have just drawn. Take this number n, put it on the horizontal axis and look at the corresponding number $f(n)$ on the vertical axis. No matter how huge the number n was, in doing this you have turned it into something between zero and 1.

'Cool trick, bro, but what can I do with this smaller number?'
Well, this is where you need the duct tape and the cardboard. You
have to make two cardboard boxes of the same height, with a total
combined volume equal to 1.

Box A must have a volume which is equal to Vol(A) = $f(n)$, that
number between zero and 1 you have just created using your
self-made function, which thus automatically implies that box B
should have a volume equal to Vol(B) = $1 - f(n)$.

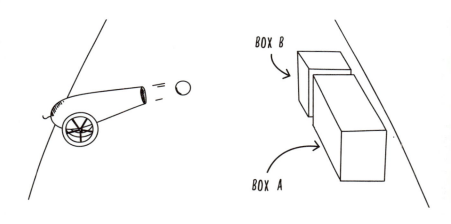

Once you have lined up these boxes, you can fire the pitching
machine at them. Just like in a baseball batting cage, you can't
control the direction under which this ball leaves the cannon, and
this is where randomness enters the picture. So you cannot know
in advance in which box the ball will end up, but you do know
that the bigger the box, the more likely it is that the ball ends up
in that box. For instance, if $f(n)$ = 0.5, both boxes are equally big
and therefore equally likely to be the recipient in which the ball
ends up. But if $f(n)$ = 0.9, box A is nine times as big as box B, which
means that there is a 90% chance that the ball will be fired into box
A. In any case, the behaviour of the ball fixes our strategy:

1. If the ball ends up in box A, you stay with your initial choice.

2. If the ball ends up in box B, you choose the *other* piece of paper.

The amazing thing is that this guarantees you a winning strategy which will work *more* than 50% of the time. Here is why: let's call the numbers on the pieces of paper n_b and n_s, with 'b' for bigger and 's' for smaller. So in order to win the Game of Googol, you should end up with the slip of paper which has the number n_b written on it. Unless you draw the one which says 'good for one free lunch': it will not win you the game, but free lunches are delicious consolation prizes. Anyway, the probability of winning the game is equal to the sum of two numbers, as there are two ways to end up with n_b. Either you choose the correct piece of paper from the very beginning and you correctly stick with it (if it ain't broke, don't fix it), or you start off with the wrong slip of paper but you decide to change. In a formula: your chance of winning the Game of Googol is equal to

$$P(n_b) \times P(\text{stick with choice}) + P(n_s) \times P(\text{choose other card}) .$$

The numbers $P(n_b)$ and $P(n_s)$ are both equal to 0.5, as you start by *randomly* choosing a piece of paper, so there is a 50% probability you will pick the bigger (or the smaller) number at the start. But that is not enough, because we have to take the strategy into account, one that says that you *only* stick to your choice if the ball ends up in box A. Vice versa, you only *change* if the ball ends up in box B. Now suppose you flipped the correct piece of paper, which reads $n = n_b$. You obviously want to stick to your (correct) choice now, which means that the ball should be launched into box A. You can't control the pitching machine — unless you've got some really badass Jedi mind power — but you know the probability with which this will happen: it is the volume of box A, which means that $P(\text{stick with choice}) = f(n_b)$. On the other hand, imagine you started by selecting the piece of paper which reads $n = n_s$. This time, you should really hope the ball will end up in box B, because this means we can choose the other piece of paper. Once again, this is beyond your control, but we do know how likely it is that this will happen: the probability is given by the volume of box B, so P(choose other card) $= 1 - f(n_s)$.

Plugging these numbers into our formula, we obtain the following probability of you winning the Game of Googol:

$$P(\text{winning}) = \frac{1}{2}f(n_b) + \frac{1}{2}(1 - f(n_s))$$

$$= \underbrace{\frac{1}{2}}_{\text{duh}} + \underbrace{\frac{1}{2}(f(n_b) - f(n_s))}_{\text{Wackiness!}}$$

This is very weird, as it implies that P(winning) is more than 50%. Indeed, the first number on the right is what you'd reasonably expect: this is the flip-of-the-coin piece of the equation, which gives you 50%. But the second number on the right *increases* this probability, turning it into something that is bigger than just 50%. 'Wait a minute, are you sure this is a positive number?' Well, yes, at least if you did your homework correctly. Remember that the function you drew had to be strictly increasing? This is what we use here: since n_b is bigger than n_s, we also have that $f(n_b) > f(n_s)$, so that their difference becomes a positive number. The only 'problem' is that you don't know in advance how much bigger than 50% your chance of winning is: it's definitely more, but you just can't tell how much more.[77] Then again, professional poker players — and people who are about to be beheaded by a guillotine — will confirm that even the slightest of edges is still better than not having an edge at all.

---

77   This is the opposite of 'rubbing salt in the wound', which is when it definitely gets worse, but you don't know how much worse. Well, unless we're talking about someone who started bleeding because of severe saline deficiency.

Recommended listening

Artist	Song title
Nils Frahm	#2
Faith No More	Separation Anxiety
The Smiths	You Just Haven't Earned It Yet, Baby
Trentemøller	Always Something Better
At the Drive-In	Non-zero Possibility
Satanic Surfers	It's My Decision
Caribou	Found out
Venom	Don't Burn the Witch
Guns N' Roses	Sweet Child o' Mine
Soulwax	E Talking

6

Space To Explore

> *Research is what I am doing*
> *when I don't know what I am doing.*
> (Wernher von Braun)

You should really read this chapter if ...

- you want to know what type of student you were during a maths exam.

- you want to turn a pile of batteries into an open problem in mathematics.

- you still think that mathematicians attending a conference are going on 'yet another holiday'.

- you wonder what maths has to say about the difference between doing something yourself and checking whether someone else did a good job.

6.1 Special abilities

Ah, the mathematicians. Often the subject of lighthearted ridicule, although the jokes about them do not always seem to make sense to me. I once overheard someone saying, 'Mathematicians really are the IKEA furniture amongst the scientists.' This was one of the oddest remarks I have ever heard, because the difference between a mathematician and a piece of IKEA furniture is actually rather obvious.

I mean, one always lacks a few screws, but can be safely put in a corner when you throw a party in your living room as it always seems to blend in nicely with the wallpaper. And the other can be bought in a Swedish retail store.[78]

A huge source of inspiration for these jokes about mathematicians turns out to be a persistent cliché: their apparent social ineptitude. To be honest, generally speaking they are not party animals of the mouth foaming type, but they have definitely got enough game to enjoy the hunt. My own experience has taught me that the alternative can sometimes be even worse: I often prefer *not* being sociable to facing the awkward silence which ensues my attempts at being chatty. For instance, at the party of a couple of friends who have bought a house, one of these occasions on which I typically find myself desperately trying to recall names of newborn babies, or at the dining table, spending a stiff first night out with potential parents-in-law, there is always at least one person who ignites the following dreaded, short-lived conversation: 'So, what do you do for a living?' This always has me spinning around my axis, hoping to see an addressee, followed up by a reluctant 'Me? I am a mathematician.'

This may all seem like a perfectly harmless invitation for conversation to you, but it is nothing less than the socialising mathematician's nightmare. I either get a puzzled face, blankly staring at me as if the owner needs more time to decide whether that really qualifies as a proper job, followed by a slick 'Oh, I see, very interesting!', smoothed to perfection with a spoonful of Irony Oil, and two eyes nervously scanning the place for the nearest exit, or an overly enthusiastic 'Ah, so we are colleagues! Which problem have you been working on lately?' The former scenario results in me trying to do what most drivers do with a reindeer caught in the headlights — steer away from the subject — the latter in *me*

---

78   On the other hand, computer scientists genuinely are the garden furniture amongst the scientists: they may come out when it's sunny for a few days in a row, but the rest of the year they are safely locked away in a dark shed.

scanning the room for the nearest exit. Let's just say that I meet enough mathematicians at work, no need to do so outside my natural habitat.

Not that I do not enjoy talking about mathematics, I even decided to write a book about it, it's just that people's attitude towards mathematicians is completely polarised. For some strange reason, meeting a maths-whizz seems to evoke strong reactions, ranging from scientophobic laughter to an almost pious form of admiration — as if we are the Chosen Ones, spending our devout lives staring at codes that need to be cracked. When someone tells me she is very good at reciting sentences backwards, I feel absolutely no urge to apologise for the fact that I am not (although I definitely would if there were a palindrome for 'I'm sorry'). But when I tell people that I am a mathematician, things seem to be different: 'Oh, mathematics. Sorry, I have never been any good at it.' Underlined with that compassionate look I personally reserve for special occasions only, like a blind colleague telling me his wife left him when she heard he had testicular cancer, and who ran over his dog on the way to her 21-year-old lover.

So if you are of the admiring type: do not waste your breath, there are better things to do with that. Like whistling loudly when you see a woman in a car who is about to hit the dog of a blind man with testicular cancer. I do think that some people are better than others at deductive reasoning and analytic thinking, but being capable of making mental leaps over massive gaping gorges — such as, say, from the latest winter jacket fashion to the colour blindness of polar bears (don't worry, this mental leap is only a few pages wide, see a bit further on in this chapter) — is by no means a guarantee of a fruitful career in science.

Definition 39.

Polar bear: *a big carnivorous white mammal living within the Arctic Circle, named after the 'polar coordinate system' — an alternative for the classical Cartesian coordinates. It is so called because no matter from what angle you are looking at this predator, it is always better to encircle it in as wide a radius as possible.*

Being good at maths is not merely a matter of talent, for the simple reason that it is also about skill. One that needs be developed, trained, practised, fed and nourished — preferably by a dedicated pro. You will not hear me saying that there is something fundamentally wrong with our educational system, but if I could have received one lousy euro from everyone who ever told me that they do not like mathematics because of a poor teacher at school, I could have been a rich one myself. I cannot really complain though, since I had excellent teachers,[79] so I am just trying to pass it along.

---

79   Mr Maes, if you are reading this: from the bottom of my heart, thanks!

I actually find this the best thing about teaching maths myself: having the opportunity to pass on both my knowledge *and* passion to students, without whom my life at university would be pretty pointless. Undergraduates are interesting creatures — dancing on that wobbly slackline anchored between formidable freedom and inevitable exams — and I embrace any chance I get to observe them at close quarters. One of the best occasions to do so is during exam supervision, an otherwise mindnumbingly dull task which barely beats 'watching paint dry' on the Universal Scale of Boredom.

The Universal Scale of Boredom:

1. Supervising students working on their final exam for the course 'Visually Establishing the Hardening of Slow Drying White Paint in Extremely Cold Conditions'.

2. Playing chess with dolphins. Yes, they are definitely clever enough, but they have pectoral fins, so they can't move any of the pieces.

3. Watching paint dry.

4. Supervising all other exams.

5. Solving binary Sudoku puzzles.

BINARY SUDOKU

0	
1	

During the many hours of exam supervision I have done so far during my career, most of which were spent shamelessly observing students tackling algebra questions, I did notice that some of them have rather special abilities. Here is an overview of the most fascinating species of students:

The Time-benders: these students seem to be capable of bending time, although rarely to their own advantage. When you observe them *during* the exam, you never see them loitering; they always seem to be completely immersed in their tasks, hastily scribbling on their exam scripts as if they are afraid that their knowledge may evaporate into thin air. More often than not in messy handwriting which leaves so much space for interpretation that if you gave their exam copy to a pharmacist, you'd probably end up with enough drugs to sedate the Balkans. But there is a catch: their answer sheets rarely contain more than half an answer to just one of the many questions, which leaves the examiner with that uncomfortable feeling that they lost the rest of the answer sheets somewhere on the way from the auditorium to their office. The truth is of course that these students only had time for half a question, *as they bend time.*

The Rubberfaces: this is the type of student you'd expect to see in a black suit and matching hat, with a white painted face, carrying a basket and picking imaginary plums. They are true mime artists, housing a mind that had different plans for the future. Observing these students is funny and entertaining, as you can deduce their thought processes and progress during the exam — or, as is often the case, lack thereof — from their facial behaviour. They surely know how to handle the complete spectrum: from pure anguish over Archimedesque moments of insight to the look on their face when they raise their finger and ask you whether they can go to the toilet (I once had the impression that the plums were not that imaginary), it all comes with a powerful visualisation.

The Non-Sensei: these students are also true artists, but in a sense which transcends reality. Unlike the time-benders, non-sensei always hand in exam scripts which contain more pages than reasonably expected. The surprising thing, however, is that their answers are a bizarre mixture of facts, ideas and shreds of course material which makes absolutely no sense. Just like drunk wombats in a book about maths, but then even less sense.

To outsiders, their exam scripts must look like exemplary answer sheets, containing the right amount of formulas and mathematical jargon, but these students have the unique — and, let us be realistic, pretty useless — ability to blend a whole lot of true facts and correct pieces of information into an answer which lies so far from the truth that you need an ultra long-haul flight to get from the question to their answer. '*Using Fermat's theorem, we may conclude that the first derivative of the cosine function is a periodic vector space with values between plus and modular infinity*': I suppose it does have a certain ring to it. The Saturnian ring of an encrypted message from outer space, that is.

The Shopkeepers: these are fairly easy to recognise. At the start of the exam, when the rest of the students are already digging into the questions like police detectives into a suspect's garden — dutifully, but slightly worried they may find something upsetting — the shopkeepers are still busy setting up their workspace. Chocolate bars, at least four pieces of fruit, a plastic box containing anything from a simple pasta salad to a full-grown gastronomical meal which leads to guaranteed steamy sex when served with red candles and smooth jazz — provided you use enough French words — a volume of water which allows you to keep a beached whale alive for a few hours, a muesli bar whose nutritional value is only trumped by the number of decibels it will produce when its owner is trying to tear the package open, a box of pens with which you can easily reproduce Shakespeare's oeuvre — a different colour for every vowel — and a body of scrap paper that used to belong to the kind of forest in which unprepared hikers die from starvation.

These students are rarely prepared for an exam — as they spent most of their time at the supermarket — but they are definitely prepared for Armageddon. Then again, some say that these are synonyms.

Definition 40.

Pharmacists: *a collective noun to denote human beings who have spent at least five years of their life mastering the dark art of deciphering doctors' prescriptions (see also: doctors). Personally, I find it very odd that no one ever thought of showing them the Rosetta Stone.*

6.2 Surely everything's been done already

Teaching at a university, a college or a high school, being a stockbroker in a financial institution or working as a business analyst in a company: this more or less exhausts the list of jobs most people have in mind when they think of mathematicians. But only a few people seem to realise that there are lots of professional mathematicians out there[80] who are doing research. Actual, current research. Cutting edge, and with a piece — too large, some suggest — of the federal science budget. Because for some strange reason, a common misconception about maths is that everything has already been done in our area of expertise.

When it comes to swallowing strange objects (light bulbs, sharp pointy objects of all sizes and shapes or a full-blown Cessna), walking on things that are not really meant to be walked upon (burning ropes, hot coals, hungry alligators, pavements on London's shopping streets during Christmas time) or doing plainly weird shit (having coral implants in the forehead, eating three of the world's hottest chilli peppers in 22 seconds

---

[80]   Well, given the fact that they spend most of their time behind a desk staring at abstract symbols: 'in there'.

or throwing yourself out of an aeroplane while solving a Rubik's Cube), you have a point: all these things have been done before. And if it weren't for its censorship policy, we'd probably have plenty of YouTube videos showing five-year-old Chinese kids doing it all blindfolded, balancing seven bowling balls and a microwave oven on top of their heads, while reciting the decimals of π backwards.[81] But when it comes to mathematical problems: rest assured, there is plenty of homework left.

Definition 41.

Microwave oven: *a kitchen appliance used by people to heat or reheat food. This is done as follows: the food is put into the microwave and heated to a temperature which is so incredibly high that whatever comes out of the oven is too hot to be consumed right away. Touching the food at this point leads to a burnt tongue and frenetic hand flapping in front of the mouth. The food is then left alone for a few moments, until it is somehow edible. This phase only lasts for a few nanoseconds tough, after which it starts cooling down so quickly that the whole procedure has to be repeated from the very beginning.*

The thing is that most people believe that all unsolved problems in maths necessarily deal with a question involving at least 17 words which would either earn you a monster Scrabble score or a maths phobia attack. This is definitely true for some problems — you can't tackle the Hodge conjecture if you did not grow up with the cohomology classes of complex subvarieties of a non-singular complex projective manifold — but some of the open problems are surprisingly easy to explain. For instance, imagine that you have a flashlight which requires two batteries to work, and a pile of eight batteries containing four empty ones. You cannot discern the empty ones from the full ones, but you can test them using the flashlight.

---

81   If the last part of the sentence made you chuckle: that is the left side of your brain talking, and it is right.

What is the *minimal* number of times you have to put two batteries into your flashlight before it will most *definitely* shine? So in a sense, we are looking for the worst case scenario: no such thing as luck involved here.

<*Puzzle alert*>

I obviously don't know which number you found, but you can safely switch on the light with seven trials only. First of all, group the eight batteries into three piles (the French-speaking readers are probably very confused at this point): 8 = 3 + 3 + 2. Now test all possible combinations in the two larger groups: this means that you will need six trials (*AB*, *AC* and *BC* for both piles). At this point, the worst case scenario is darkness (still) imprisoning you (all you can see, absolute horror). But this then allows you to conclude that the remaining two untested batteries are both non-empty, so you can safely put these into the flashlight and switch it on. Just to make sure, you still need to prove that it can never work with six trials only — but this is left for the reader as an exercise.[82]

A normal person's reaction to the problem described above might be to get rid of the empty batteries, but a mathematician sees it as a standing invitation to generalise the problem and focus on what I will hereby coin as the *flashlight number*. Imagine a flashlight requiring *c* non-empty batteries to work, and a pile of *n* batteries in total containing *g* non-empty ones. While your partner is out — buying a pack of new batteries (or really embracing the dark and decapitating a pile of hamsters) — you define the flashlight number $F(n, g, c)$ as the minimal number of trials needed to make that flashlight work. As far as I am aware, this number is not known in general. I am not saying you will change the world if you can calculate, and prove, the value for arbitrary (n, g, c) — if

---

82   Just kidding. Actually, this is one of the most hateful sentences appearing in
      maths textbooks in my opinion. It's the slightly less condescending version of
      the phrase 'It is trivial to see this.'

you really want to make a difference, why not invent garbage bags that take themselves out when they're full — but it would be a solution for an open combinatorial problem in maths. Then again, maybe flashlights simply do not arouse you, but that problem is more easily solved: you can replace the 'a' with an 'e' in that case.

I haven't got the foggiest whether we will ever come to that point where we can proclaim that 'everything is now known in mathematics', but the idea itself scares me a bit: dropping a whole bunch of jobless mathematical researchers on the labour market makes me think of putting that little box of toothpicks on the table when you are hosting a dinner party. Every once in a while someone decides to choose one and put it to use, but most of the times it ends up being an odd piece of decoration. I am not claiming that mathematicians can't do anything *but* mathematics — like I said, most of us end up in IT, the financial sector, consultancy companies and schools — but my own experience has taught me that once our mathematical brain takes control in daily life situations, things often go in a non-conventional direction. For instance, I once worked as a barman in a pub.

Me: 'That will be 5 euro 35, sir.'

Customer: 'Make it 6, if you will.'

Me: 'Erhm, okay, that will be 6 euro 35 then.'

Not really a recipe for success, right? Speaking of which, that's another thing I once tried. 'You like food and you enjoy writing,' I told myself, 'so why not start a recipe book?'

Pancake recipe

Ingredients:

- the first perfect number of eggs

- 1,234 grams of flour in base 6

- 37 litres of milk (modulo 36)

- salt (s > 0)

Method:

Pour flour and salt into a mixing bowl (or any other container which is topologically equivalent to a disc). Choose a coordinate system with axial symmetry, and create a global minimum at the origin. Crack the eggs into this paraboloid and add one sixth of the milk. Take a kitchen utensil shaped like a manifold with sufficiently high genus and start whisking from the centre, following a logarithmic spiral until all the flour is drawn into the eggs. To finish the batter: pour in the remaining milk (make sure the flux is constant) and keep whisking until the consistency is isomorphic to single cream. Heat a bounded surface with a handle over moderate heat, ladle some batter into it and tilt the pan in a sharp angle in order to transform the batter homotopically into a low eccentricity ellipse. Let it now cook, undisturbed, until you have recited the digits of the 14th Mersenne prime number. Once that is done, hold the handle and ease a fish slice under the first pancake. Flip it, unless you want to make Möbius pancakes: in that case just leave it be. Wait another 30 seconds and then consider a Euclidean transformation from the pan to a plate. The rest of the method now proceeds by induction on n, the number of pancakes.

MÖBIUS PANCAKE

I guess I was lucky enough to find my thing at university, where I started studying maths, and to become the proud holder of a PhD in mathematics. I still vividly remember the day after my PhD defence: I woke up in a state of exhilaration, but I kept wondering why I was now being referred to as a *doctor* in mathematics — that just didn't make sense to me. So I decided to do an experiment: I went to a local pharmacy just around the corner, with a hand-written page of my mathematical research (a truckload of integrals in higher dimension). To my great surprise, I went home with two boxes that day. The pharmacist briefly eyed my paper, and gave me a box of Dafalgan and a giant box of condoms. The Dafalgan made sense to me, mathematics seems to have this unique tendency to give some people an instant headache. When I asked him about the condoms, he just said, 'I really think you should start integrating these into your life instead.'

Regardless of this question whether the mathematical puzzle is indeed finite or not, there is still much to do at this point. As a matter of fact, there is a rich history of to-do lists in mathematics. The most famous one is the list containing Hilbert's 23 Problems. It was compiled in 1900 by David Hilbert, whom we met earlier in the context of the infinite hotel, and presented at the International Congress of Mathematicians in Paris. Most of these problems have been resolved in the meantime, if not at least partially, but a few still remain unanswered. Hilbert's problems cover a wide variety of topics, and were designed to serve as a source of inspiration which could potentially lead to fruitful advancements in several mathematical disciplines.

David Hilbert

As I find this a magnificent guiding philosophy, I have taken the liberty of compiling the Hilbert Programme in Social Engineering. Having attended a fair share of receptions and formal banquets in my life — for instance at maths conferences — I would be the first one to embrace any attempt at solving the list of problems below.

1. **The Seated Dinner Table Timing Dilemma:** this problem has to do with the optimal timing for entering the dining hall where a formal banquet is to take place. On these occasions, one often faces the following issue: entering too early

enhances the risk of choosing a table which attracts the people you want to avoid at all costs. Coming in too late, on the other hand, means that there are fewer seats available and this often forces you and your company to break up and sit at different tables. If there are n people in total and ρ round tables, determine the optimal time $t(n, \rho, \varphi)$ to enter the room if you are out with φ friends.

2.   **The Public Kissing Problem:** does there exist a universal answer to the question of how many times one should kiss a person, and which cheek to start with? The answer should be a number $N(d, n, \sigma)$, which depends on a parameter d (the degrees of separation between you and the kissee), a number n (keeping track of how many times you have met that person before) and another constant σ depending on the receiver's social context. Also, while we are at it: where exactly lies that fine line separating a polite kiss on the cheek from one that is categorised as 'seductively close to the mouth' (yet another kissing problem)?

3.   **The Catered Reception Issue:** what is the absolute minimal amount of time you have to engage in conversation with someone who bores you to death before you can politely walk away? Also, given the surface area A_v of the venue, the number n_s of wait staff serving the audience and the number n_p of people in the room, is there an algorithm which tells you where to position yourself if you want to maximise the number $S(A_v, n_s, n_p)$ of snacks you will get? You may assume here that the wait staff flux is constant.

4.   **Public Urinal Ponderings:** if the number of urinals in a public toilet lies roughly between three and six, and if there is no immediate queue, men are confronted with the problem of having to choose where to stand. An unwritten rule says that one should avoid standing directly next to someone, and if the last urinal in line borders the door, you have to make sure that the next person will not be forced to take this one.

An optimal strategy, depending on the number of toilets and people peeing, seems to be missing.

5. **The Private Conversation Equilibrium:** if you have ever been invited to a birthday party held in a friend's living room — the kind of party where everyone knows someone but no one knows everybody — you might be familiar with this problem. As long as the number of people sitting around the coffee table is too small (four or five), social etiquette seems to dictate that you have to participate in an overall discussion: it's either talking, or nodding. Starting a private conversation with your neighbour seems impossible at that point, and sometimes even frowned upon, whereas this becomes perfectly normal if there are more people. This leads us to the question of the equilibrium: what is the lowest number of people required?

Some of the standing problems mathematicians are working on are so desperately in need of an answer that huge financial rewards have been put on their heads. The 'Millennium Prize Problems' are seven problems in mathematics that were chosen by the Clay Mathematics Institute in 2000, and as such qualify as the 21st-century analogue of Hilbert's problems. At the time of writing, six of them are still unanswered. A correct solution earns you a whopping one million US dollar prize — in sharp contrast to the Hilbert Problems, which earned you admiration from the mathematical community and Hilbert in particular (I know, quite the incentive — right?).

One problem, the Poincaré conjecture, was solved by Grigori Perelman around 2003, but he declined both the award and the money. Just in case you did not get the previous paragraph: he *declined* a million dollar prize, and just went back to doing research in mathematics. This somehow makes me think of that corrupted question which inevitably seems to pop up when a romantic relationship with someone slyly crosses the border between the Lands of Loosely Holding Hands and Boldly Entering

The Bathroom To Do The Number Two When Your Partner Is Having A Shower: 'Darling, would you still love me if I were to end up in a wheelchair one day?' Because by turning down his monetary rewards, it seems like Perelman gave the Queen of the Sciences a correct answer when she asked him the mathematical equivalent of this question: 'Darling, would you still spend time with me if they'd offer you so much cash that you could do anything you want for the rest of your life — like travel around the world to sit on a chair for a few hours, merely staring pensively at what most people consider to be abstract bollocks?'[83]

Explaining in full detail what Perelman did is not an easy task, and I don't mean that in a condescending manner. Let us just put it as it is: some things are harder to explain than others. For instance, if I have to choose between explaining why polar bears are colour blind, or why the doctor asked us to blow on our wrists when he wanted to see our testicles during the annual medical checkup at secondary school, I'd surely know which one to pick. The latter had an immense impact on my sex life by the way: every time a woman is going down on me, reverse logic forces me to check my wrists.

Definition 42.

Doctors: *a collective noun for human beings who have spent at least seven years of their life learning to write messages in dark illegible codes. So far, the only creatures in our galaxy known to be able to decipher their notes are also found on planet Earth. Interstellar theoretical sociologists are convinced that this is not just a coincidence: they believe that for each doctor, there exists a unique anti-doctor (see also: pharmacist). To prove these bold claims, researchers are trying to recreate the beginning of the universe, because they claim that during the first few seconds of our existence, doctors and anti-doctors existed as a singlet. This symmetry then broke down during the cooling phase of the universe.*

---

83   Taking your time at the finest museums of contemporary art.

6.3 The art of not solving a problem

I guess it comes as no surprise that most problems on the Millennium Prize list require a fair bit of knowledge even to understand what they are dealing with, but the so-called **P** versus **NP** problem is slightly more down-to-earth. To **P** or **N**(ot to)**P**, what is this question in the first place?

Let's assume that someone did something nice for you, like cooking you a meal, ironing your clothes or wrapping that soccer ball you bought for your nephew's birthday in colourful paper. Imagine also that your well-meant feedback — sprinkled with the gentlest of suggestions for improvement — is taken for harsh criticism and therefore met by a poisonous 'Why not do it yourself then next time?' If at that point you don't really feel like accepting the challenge, you are probably familiar with the idea that it is sometimes easier to judge whether something is done well than to actually do it.[84]

The **P** versus **NP** problem is a major unsolved issue in computer science, which essentially boils down to what I have just described: informally, it asks whether any problem whose solution can be quickly *verified* by a computer can also be quickly *solved* by that computer.

Whereas politicians are trained to be as vague as possible about things, the exact opposite holds true for mathematicians: they must quantify and formalise everything. The branch of mathematics called calculus for instance — bringing you quality nightmares involving limits, derivatives and integrals since the mid-17th century — can be seen as the result of people trying to quantify the notions 'infinitely big' and 'infinitesimally small'. I suppose this is why people often say, and I have the impression it is not always meant as a compliment, that mathematicians

---

84   A crucial exception to this rule is grading exams: I'd rather *solve* a maths exam than *grade* the whole bunch.

cannot think except in numbers and figures. I would say the world sometimes needs even more things to be quantified: I have been called 'a cute guy' on several occasions and I am still not sure how to interpret this notion.

Definition 43.

Politicians: *as a great politician once said: 'Someone who often — which, as a temporal adverb, does not really say that much because it gives no clue about the actual frequency of what is about to follow — produces the kind of sentences to convey a message which are characterised by long enumerations, references to purposely unspecified sources and nested structures — for which the speaker uses intentional pauses, air quotes and miming techniques which cover the point they intended to make under a thick layer of irrelevant data, not unlike the crust of the Earth covering crucial lessons about the history of mankind and its origin under a collection of rock strata and volcanic residue — is a natural-born politician, having that unique ability to leave a confused audience behind, wondering whether they missed the point, or there wasn't one in the first place.'*

The term 'quickly' which appears in the description of the **P** versus **NP** problem is no exception: in order to make sense of it, it needs to be quantified. As a matter of fact, the clue is in the name: the letter **P** stands for *polynomial* and represents the 'speed' by which the solution can be obtained. Note that the actual speed of your computer is irrelevant here, it's just a theoretical measure. This is why I use the phrase 'the number of steps' in what follows. Technically speaking, I should add that it concerns decision problems here:[85] these can be posed as a yes-no question about the input. For instance: given a number n (input), is this number prime (output)? If you are not convinced decision problems can be

---

85   Well, I should rather say that it concerns 'decision problems fed into a deterministic Turing machine'. But just like when you start clicking links in that one Wikipedia article you had to look up, this would just lead us too far astray.

hard, I bet you never had a partner standing in front of the closet containing her party dresses. On a side note: be careful when she has narrowed down her problem to two dresses, and casually asks you whether you prefer the dress on the left, or the one on the right. This is a ROR-question (if this doesn't make sense to you, see Definition 13).

Suppose, for instance, that a computer has to pick the largest number in a given set of numbers (as a decision problem: given an arbitrary number in the set, verify whether this is the biggest number in that set). There is an easy algorithm to solve this problem. Start from the instruction.

- Retain the first number in the set

and then go through a loop with the instructions

- check the next number in the set

- retain this number if it is bigger, otherwise ignore it

until the end of the set is reached. Serial monogamists should be familiar with this, as this is their dating strategy. Despite this being a rather trivial example, it allows me to explain why computer scientists call it **P**: if the set contains n elements, then a computer will need n steps to solve this problem before it can go back to business as usual — undermining your self-confidence, asking you whether you are *really* sure you want to proceed.

In complexity theory, one then says that our example is a problem of order n, compactly denoted by $\mathcal{O}(n)$. The more elements, the more difficult, but in a linear way. Or in mathematical lingo: the size of the set and the degree of difficulty are directly proportional: if the set becomes twice as big, it will take twice as long to solve the problem.

Definition 44.

Complexity theory: *a branch of computer science in which compu-*
tational problems are classified according to their inherent difficulty.
So complexity theorists are like professional wine tasters who spit the
liquid they evaluated back into a bucket: they do not solve problems,
but at least they can tell you how hard they are.

Slightly more difficult would be trying to match two people in
your group of friends. Not least because for some strange reason,
people always seem to know when they're being set up on a date. I
mean, when we leave to go on holiday our brain can't remember
whether we turned off the oven, but it can sniff out a blind date
as if it were nothing. Anyway, suppose there are n women and n
men, and you are trying to fix the most compatible match. There
are n^2 possible couples, as you can pair up each of the n women
with each of the n men, and these need to be compared somehow
(I told you, there is still a lot to do in the Quantification Department). This time, complexity theorists will say that we are dealing
with a problem of order $\mathcal{O}(n^2)$. This is still polynomial, but in a
quadratic way: the number n is to be squared. Indeed, doubling
the size of the group leads to four times as much work.

The majority of real-life problems are obviously more complicated, but as long as the solution requires a number of steps which
grows as a power of n (like n^{42} or n^{666}), it is still **P**. Computationally
speaking, these (decision) problems are 'easy' for a computer to
solve, as the algorithm is 'polynomial'. Note that most real-world
P-problems are $\mathcal{O}(n^2)$ or $\mathcal{O}(n^3)$, you will rarely find higher powers.
When it comes to big exponents, nature seems to have spent all of
its credits in the Quantum & Cosmology Department.

On the other side of the equation,[86] we have the class of problems labelled as **NP**. These problems are either shown to be hard
to solve, or it is still not known whether there exists an easy solu-

---

86   Well, there are actually loads of different complexity classes. Understanding
      them all is a problem which deserves a class of its own.

tion (note that **NP** does *not* stand for non-polynomial, this is a common misconception[87]). But, if someone claims that he or she has found a solution (you can always hope for that nightly visit from the Hindu goddess Namakkal), it is easy to verify this. To be more precise, it means that there exists a polynomial proof to check that the alleged solution is indeed correct.

Suppose, you are given the set $\{-2, -3, 15, 14, 7, -10\}$. Is there a way to add some of these numbers (you do not necessarily have to use them all), so that you end up with zero? If you tried it yourself, you probably noticed that $\{-2, -3, 15, -10\}$ does exactly that: adding these numbers gives zero. Verifying this is simple for a computer, even if the set contains a lot more numbers: if the suggested solution consists of n numbers, the computer will have to perform $(n - 1)$ additions, which is a polynomial algorithm of order $\mathcal{O}(n)$. However, and this is where the excrement hits the oscillating air current distribution device: solving this problem is way more difficult. In a sense, the only option is to try *all possible* combinations of *any* number of elements.

For example, we have started with six numbers in the example above, so in a worst case scenario we have to try all possible combinations involving two, three, four, five and six elements. Without going into details here, this will lead to an *exponential* complexity, not a polynomial one (there are several ways to tackle this problem, some of which are 'slightly faster' than the others, but these algorithms are all of exponential order). This means that if your set contains n elements, it may cost you roughly 2^n steps to find a solution in this particular example.

To get an idea of precisely how bad exponential behaviour actually is compared to polynomial behaviour, just look at the table below. The number n in the left column represents the 'size' of the problem (for instance, the number of elements in the set under

---

87   It stands for non-deterministic polynomial; see the literature if this is the kind of stuff you want to know more about.

consideration), whereas the columns labelled $\mathcal{O}(n^5)$ and $\mathcal{O}(2^n)$ respectively contain numbers representing the number of steps it takes to *verify* a given solution versus the number of steps it takes to *construct* a solution[88]:

n	Verify $\mathcal{O}(n^5)$	Solve $\mathcal{O}(2^n)$
10	100,000	1,024
25	9,765,625	33,554,432
50	312,500,000	1,125,899,906,842,624
100	10,000,000,000	1,267,650,600,228,229,40 1,496,703,205,376
500	31,250,000,000,000	bleep

Suppose that a set has 100 elements and that the computer can handle one million operations per second. It then follows from the table above that the *verification* of the solution will cost you a little less than three hours, whereas *solving* the problem may cost more than 10^{30} seconds. A gargantuan number, approximately equal to $40, 000, 000, 000, 000, 000, 000, 000$ years. It's hard to get an idea of how long this actually is, but let's say that the estimated age of our universe doesn't even come close to this number!

Computer says 'no'

So why is this problem so important? Apart from being a beautiful problem in itself — not only connecting mathematics and computer science, but even extending into the realm of philosophy, probing the very limits of human knowledge — it mainly attracts interest because of the consequences it might have. If there is one thing I am trying to avoid — apart from low flying drunk pianos — it's definitely thinking too much about things that *might*

---

88   Just to be sure, n^5 and 2^n are merely examples here; I could have chosen n^{37} and 42^n too. This will obviously change the numbers, but not the overall conclusion.

be. And yet, in the context of the **P** versus **NP** problem, it seems to be worth doing so. Although most mathematicians and computer scientists believe the opposite statement is more likely to be true, a proof for **P** = **NP** could lead to efficient ways of solving **NP** problems, as it would imply that the apparently hard problems actually have relatively easy solutions.

This has advantages and disadvantages: modern security applications are based on algorithms which would suddenly become easier to crack than an egg interrogated by a Gestapo officer. So this could spell the end of secure financial transactions over the internet. On the other hand, some of the most challenging problems in operational research and logistics — such as the travelling salesman problem — would suddenly become tractable, spurring considerable advances in many other branches of science and technology. So the world hunger problem will not be solved if **P** and **NP** turn out to be equal (well, unless it stands for the giant pea from chapter 2), but if someone else does solve that problem, at least the solution will go around considerably quicker.

It could even have consequences for mathematics itself, in the sense that computers could suddenly be able to *generate* proofs for certain theorems: in goes the list of axioms, out comes a list of theorems. 'How do you mean?' I hear you thinking. 'Like robots spewing out papers?' Yes indeed, something along those (coded) lines. You may think this sounds a bit scary, but I am afraid it's already too late to be concerned about that: nowadays, researchers at universities have to meet demanding criteria concerning their publication output and this does not always encourage (young) people to pursue a path if there is no foreseeable publication, preferably in the plural. Although this is a very personal remark, coloured in a dark hue by my burn-out experience, I do think this poses a slumbering threat to science in general: **P** versus **NP**, where the capital **P** stands for publication pressure.

Definition 45.

The travelling salesman problem (TSP): *most people have the following problem in mind when they are talking about the TSP: 'Given a list of cities and the distances between each pair of cities, what is the shortest possible route that visits each city exactly once and returns to the origin city?' This is a common example of mistaken identity, though, as the real TSP refers to the following list of problems:*

- *Why are there always at least 17 pillows on a hotel bed, with a degree of fluffiness ranging from the stuff dreams about winter clouds are made of to pure concrete?*

- *Will my boss find out when I order* Ass Ventura: Smut Detective *from the hotel in-room entertainment system?*

- *Suppose that each time after I went to the toilet in my room, I concluded by making that typical V-shaped toilet paper fold myself. Would there be a point at which the cleaning staff spontaneously start offering me laxatives?*

- *Why does the hotel even bother putting up messages like 'for the sake of the environment, we kindly ask you to reuse your towels' if the cleaning staff take them anyway?*

- *Why does every freshly made hotel bed feel like a pair of sheets stapled to the bed frame?*

THE TRAVELLING MATHEMATICIAN'S PROBLEM

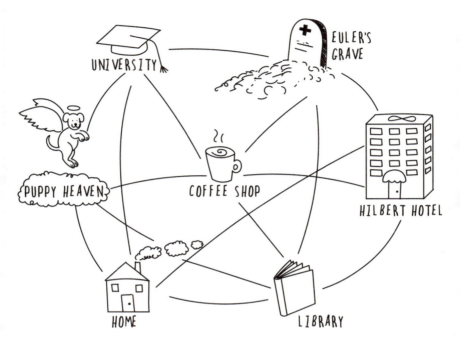

6.4 The travelling mathematician's problem

I am absolutely not claiming that every mathematician is aspiring to solve one of these main open problems; on the contrary I would say, but there are still plenty of new theorems proved every day. So how does this work? How do people invent new mathematics? As it is all about creating something original — which amounts to crossing your fingers and hoping that the muse's visiting schedule contains your name — I sometimes compare doing research with what musicians are doing. Although I have to admit that this comparison does a better job at describing a broken dream (ever since I discovered hardcore punk music, I wanted to be in a band) than the actual truth. My life as a researcher does not involve drugs, wild adventures with groupies and an ecstatic crowd shouting for more. Unless you count the coffee, the occasional wink from the cute librarian and the lukewarm applause from a handful of dozing professors and PhD students, staring at my presentation like a squirrel at a coconut ('I like it nutty, but this is too much for me').

As for a decent comparison, my best shot is probably to resort to sex. Doing research is like mental masturbation: a fair amount of paper is wasted along the way to enlightenment — even the occasional napkin, picked up during dinner — and it is not always useful for humanity, but the conclusion always justifies the effort we had to put in to reach it. And surely, it helps to read related material (papers in journals), to hear other people talk about it (at a conference, or during a seminar[89]), or to get in touch with a proverbial helping hand. But my personal experience has taught me that this seldom leads to the same satisfactory feeling as the mathematical version of a wet dream. Every once in a while, when we least expect it, it just happens. Out of nowhere. Like a lightning flash of accumulated concentration. And this outburst of proper inspiration contains enough seeds for a few brainchildren we can raise on our own, reducing the rest of our time behind the desk to cleaning up the details.

In practice, this amounts to writing a paper containing your claims, and sending it to the editor of a mathematical journal. This person then forwards it to a few other mathematicians — experts in the same field, remaining anonymous to the authors — who are to read the paper thoroughly, and to write a report containing their recommendations. *The X Factor* for people trained to find *x*, so to speak. Only if these reports are favourable — or if the authors can figure out the identity of the referee, so that they can leave a severed horse head in her bed — the paper has a chance of appearing in the next edition of the journal. In sharp contrast

---

89 Pun alert: make that a *semenar*.

to other scientific domains, like biology or physics, this phase between the submission and the actual appearance of a mathematical paper often lasts more than a year. Not least because the authors sometimes have to wait more than six months before they get the following message: 'Due to a huge backlog, we cannot consider your paper for publication.' Which is like making reservations at a restaurant and only being told at the door that the place is fully booked.

Definition 46.

Mathematical journal: *the educated version of a* Playboy *magazine: some people get really excited whenever there is a new issue, and although these journals always contain pictures conveying messages which are so much harder to describe in full sentences (things like graphs and geometric illustrations), everyone claims that they are buying them for the articles.*

This delay in the dissemination of mathematical developments probably explains the success of websites like https://arxiv.org, where researchers post unpublished and non-peer-reviewed versions of their papers, which are freely accessible. This eventually leads to the same dynamic as gossip: by the time the news finally appears, everone already knows about it and does not really care anymore. Nevertheless, as long as a paper has not officially appeared somewhere, it does not really count: researchers need their publications to have an official code (something along the lines of 'A not so short note on the inexplicable rules governing the abbreviation of journal names', J. of Rid. Math. Abbrev. **42** No. 666 (2017), pp. 37-1291) or they cannot appear on their publication record. I guess references are to scientific papers what pictures are to social life nowadays; apparently you need them, or it did not happen. Inexplicable abbreviations are not the only strange thing about mathematical journals, for some bizarre reason they also contain plenty of euphemisms.

We already learnt that 'trivial' is one of those words whose intended meaning can be as far from your interpretation of it as a child's drawing of a house from an architectural plan, but there are more examples of this:

Mathematical Phrases	True Meaning
It is believed that	I think
It is generally believed that	My friends believed me when I told them at a BBQ party
A well-known result	I have no idea about the original reference
Statistical interpretation suggests	It is probably wrong
In this paper, we restrict our attention to three special cases	The rest is just too difficult
This will be treated in an upcoming paper	We have no idea about this yet (but hands off our topic)
This is left as an exercise for the reader	Please mail me your solution

When it comes to disseminating new research ideas, the highlight of most practising mathematicians' working year is probably their annual conference schedule. Scholars are not merely encouraged, but quite simply expected to go to conferences and meetings on a regular basis, in order to deliver at least one lecture themselves and attend as many of their colleagues' presentations as possible.

Judging from the type of questions I often get from friends, it is safe to conclude that most people have absolutely no idea what happens there — 'What? Are you having a holiday again?' — so I decided to sketch a typical conference week.

Unless the symposium falls during the summer holiday season, it needs to be wedged in between regular classes. In my case, this means that I will either have to bribe one of my PhD students with a research idea (option A: ask a subordinate to take over the lecture), a colleague who does schedule planning with a bottle of sparkling wine (option B: have them rearranged) or my students with an exam question (option C: cancel the classes). In any case, the race around the clock starts off with a pang of guilt.

In an honourable attempt to reduce the travel costs — there is only so much I can do with that bench fee — I usually have an itinerary involving too many flight legs and not enough armrests. As a result, I am often so sleep deprived by the time I arrive at my destination that I look like a badly drawn version of the person on my passport picture. At least this explains why we are no longer allowed to smile when they take our picture: we all look grumpy upon arrival anyway.

The first day of the conference is usually centred around the following basic questions: where, when and how often are the coffee breaks held, will the name tags come with a ribbon (yes please) or one of those damned safety pins (I own so many punctured shirts at this point that it looks as if my closet is inhabited by a very picky clothing moth — 'breast pocket fabric, *omnomnom*') and how fast is the wifi (gone are the days when 'having a good connection' meant you could score some decent weed)? In the meantime, I am given yet another tote bag with the conference acronym — with all the bags I have accumulated over the years I can sew my own sponsored yurt — and a welcome speech by a mayor who is trying hard to give the impression that she knows why experts from all over the world have travelled to her city.

Once the ceremonial part has been taken care of, the real stuff begins. A whole week of lectures, which come in two kinds: main lectures of general interest, usually before noon, and parallel sessions focusing on niche topics, often starting after lunch. The former fall victim to the Law of Conservation of Misery (the more interesting the topic, the more incompetent the speaker and the louder the background noise generated by the uninterested people sitting next to me), the latter to Murphy's Law (the interdisciplinary lectures I absolutely don't want to miss are all scheduled during the very same time slot: the one containing my own lecture).

Conference attendants obviously come in stereotypes. First of all, there is *the interruptor* — the heckler of the bunch — who just loves cutting people off to point out that whatever they have done was done earlier, and more elegantly, more often than not by himself. In fact, if the previous sentence sounded vaguely familiar: he probably already wrote it somewhere, but better worded. Only slightly less irritating is *question-man*, who suffers from an irresistible urge to ask at least one question after every presentation. Before you accuse me of ridiculing people taking their job seriously, let me point out that his only incentive is to boast about his own expertise; if there ever was a proper polymath, he seems to be the man. The one thing that I have learnt from these people is that you should never be afraid to ask a question in public, even if you slept through the whole talk: just improvise a sentence involving a Russian surname, your favourite dimension higher than five and a few polysyllabic words. As long as you raise your intonation towards the end, you look like an absolute genius. You don't even have to be afraid that the speaker will not get your question: you can always count on the interruptor.

'Luckily enough it's not just the audience,' I hear you thinking, 'you go there for the speakers.' True, as long as it's not *an Acrobat reader*: this kind of lecturer literally projects a full PDF version of his latest paper, and tries to read it out loud within the allotted time — always at least half of what he had in mind.[90] The reading out loud is not even the worst part: as the tiny text font is illegible from a distance, it is more or less the only way to get an idea about his research. The real problem is that an Acrobat reader always gives the impression of having never seen his slides before, squinting at the screen like a drunk friend pushed on stage in a Vietnamese karaoke bar. There is a silver lining though: their lectures are perfect occasions to catch some sleep without being caught doing so. In a sense, these speakers are like turtles: unless you turn them around yourself, the only thing you will ever see is their back.

More annoying is the *time limit crosser*, often a true nightmare for the poor chair of the session in which this speaker delivers his lecture. For the readers who are not familiar with conferences and their unwritten rules: each session comes with a chair, a designated member from the audience whose thankless task it is to make sure that each lecture ends on time (this is particularly important when there are parallel sessions, so that people can make their way from one lecture room to another). On average, a speaker gets 25 minutes with an additional five minutes for questions from the audience (see: question-man). If the chair takes the job seriously — unfortunately this is not always the case — he or she will subtly let the speaker know that the 25-minute mark is approaching: with a nod of the head, using a hand gesture or standing up and slowly walking towards the stage. But in the case of the time limit crosser, the chair can do anything (fire an imaginary bazooka, mimick a slit throat, catch a white shark using two violins and a piece of raspberry pie), it simply does not work: the speaker will continue the lecture, completely undeterred. It's never that the warning signals were not perceived, it's just that the time limit crosser's oratory techniques are like a television cook's lasagne: well-prepared in advance. It usually starts with 'this is the penultimate slide' (a phrase that will apply to the next 13 slides), followed by 'there is one more thing I would like to add here' (a

---

90   Now that I think of it, they might be Time-benders with a PhD!

remark which precedes another seven-minute speech — hereby making the warning as hollow as the inside of an empty barrel of nothingness), and often ends with sheer emotional blackmail of the (still not intervening) chair, with a 'if my highly esteemed colleague would be so kind as to let me finish this sentence' (whoops, another four slides). This is usually the point where an otherwise pacific audience expresses the need for a chair armed with a whole array[91] of pointy objects.

And just when you thought that you had finally managed to get a grasp on how 'entertaining' a conference can be, there is that paragraph on the day of the social event. Every symposium comes with a built-in moment of rest: usually a half-day excursion, followed by the highly anticipated social dinner party — my source of inspiration for the aforementioned Hilbert Programme in Social Engineering. Despite this being an official day off, it usually starts with a proof by collective example. Not for a mathematical theorem, but for the fact that there must exist such a thing as 'the School Bus Syndrome'. Arm at least 20 people with a packed lunch, a name tag and a voucher for free lemonade, and you get exhilaration of the giggly kind, unforeseen wee stops and a nervous city guide trying to explain to the mildly annoyed bus driver that someone has gone missing. Not necessarily because some people deliberately wandered off during the trip to the local guy who makes traditional flutes from cured pig intestines, but because a few participants went from discussing maths on the bus to drawing arcane symbols with a stick in the gravel and got separated from the main bunch.

---

91   For the mathematicians: matrix.

THIS CAN'T BE RIGHT.
THERE'S NO PI.

The social dinner party, later that day, is where the second collective proof is delivered: pour enough alcohol into an introvert scholar, and you end up with a cross between a fire alarm, a Saharan sandstorm and a tattoo — annoyingly loud, relentless and difficult to get rid of. If you are lucky, it's a walking dinner: this means that you can easily sneak away during the ceremony marking the end of the day — a few speeches and a performance by the local Cured Pig Intestine Flute Band — but by the time you arrive at the buffet table most food has been stacked onto the plates of hungrier colleagues, clearly mistaking the concept of a shared dinner for a hurricane survival kit shopping spree.[92] At the other end of the spectrum, there is the seated dinner: less control over who your neighbours are, but at least you are being served — by a confused waiter who wonders why the guests are cracking up with laughter when negative epsilons are mentioned. Needless to say, by the end of that evening — when the mayor, who still doesn't seem to have a clue what the mathematically educated people are doing in her town, delivers the closing speech — I sometimes begin to ask myself why I didn't start the week by bribing myself, with a week-long retreat (option D: none of the above).

---

92   Now that I think of it, they might be Shopkeepers with a PhD!

One other thing that friends often want to know when I am leaving for a conference, is the topic of my lecture. Either because they find exotic titles funny — some of them do look like the result of a hot drink spilled over the Scientific Terminology Generator[93] — or simply because they are still not convinced that there is mathematical stuff left to talk about. I will not bother you with the research I have been conducting over the past few years, but I did compile a small list of mathematical subjects which have not been considered so far. A few plans for the future so to speak, once the to-do lists have been taken care of. They are all based on existing fields of research, but I have generalised them in one way or another:

- **Non-communicative geometry:** *non-commutative* geometry is the branch of mathematics studying geometry in a world where, roughly speaking, *AB* is different from *BA* (which may sound a bit bizarre to people who did not go beyond real numbers in their math class, as we all grew up with the idea that the order does not matter when you multiply two numbers). 'But surely life is a lot easier when you keep everything commutative,' you may wonder. Given my expertise, I don't believe I am qualified to answer that question, but how about asking someone suffering from dyslexia? Even something as simple as dinner is not commutative: I bet having lasagne *followed* by a plate of antipasti can get you killed in Italy. *Non-communicative* geometry is concerned with the abstract principles governing misunderstandings. One of the fundamental questions is the following: given one person's hard-line vision and a point he does not agree with, does there always exist a parallel vision which disagrees with the original one over the whole line? Moreover, one should also investigate whether all situations in which someone gets cross about another person's circular reasoning lead to orthogonal standpoints.

---

93   Go to https://pdos.csail.mit.edu/archive/ scigen/ if you feel like generating a complete scientific paper!

- **Tipology:** *topology* is a branch of mathematics concerned with the properties of objects which remain invariant under continuous deformations of space. This basically means it is the maths underlying your sex life: stretching, bending, pushing and pulling are allowed, but you cannot make extra holes. So for a topologist, circles, squares and triangles are all the same, as they can be continuously deformed into one another. This may seem irrelevant, but it does have interesting consequences: you don't need to worry if your toddler keeps trying to push the square piece into the round hole of the shape sorting puzzle as she might still become a topology expert. There are quite a few topology experts in my department at the University of Antwerp. When I told them that I was writing a book on maths and humour, they asked me not to make fun of them. I gave them seven balls and said: 'You know what, if you figure out how to stack these cubes, I promise I won't.'

Along the same flexible line of thought, there is no topological difference between a coffee mug and a doughnut: they can also be deformed into each other without breaking or gluing, in such a way that the handle of the mug becomes the opening of the donut. This begs the following question: why is there no raunchy dating show along the lines of *Ex on The Beach* and *Temptation Island*, featuring nothing but busty blonde bimbos and topologists? Because the latter do seem to adhere to the right philosophy for this kind of television: a hole is a hole.

Things that are topologically like a doughnut
(characterised by the presence of one hole)

another doughnut

a coffee cup

dear Liza's bucket

the killer's alibi

Tipology is a completely different branch of pure mathematics, dedicated to the problem of rounding off restaurant bills. Some say that the amount of money you leave your waiter should be independent of the size of the dish — unless you broke your plate — but others find that a bit of a stretch and point out that disrespectful tipping is one of the main reasons why waiters sometimes feel like they are the square peg in the watering hole. Therefore, they argue, a universal tipping formula must be developed. Preferably one that only requires utterly basic arithmetic, so that we — mathematicians — can finally enjoy a night out with friends, without having to calculate the whole thing.[94]

6.5 A colourful theorem

To conclude this chapter, let me get back to that statement on polar bears I made earlier. Being a researcher myself, I know perfectly well how long and lonely the night can be when you are being haunted by a riddle you can't wrap your head around, so here is a proof.

Theorem 3. *Polar bears are colour blind.*

Proof: Once I accepted the old adage *'quality over quantity'* as an axiom in my life — one that applies to pretty much anything but empty train seats and aeroplane armrests — I started going to quality stores for certain things as opposed to buying them in a supermarket or a retail store. This philosophy brought me to one of these big outdoor clothing and equipment stores one day, as I wanted to treat myself to a decent winter jacket.

94　I guess my frustration about forced number crunching at pubs and restaurants is like an ocean floor lobster training for the 200 metres: running deep (see also: theorem 1).

Little did I know that going there for a coat is like showing up on a first date wearing a bowler hat made of meatloaf, with a buzzing dildo on top: it was doomed from the very beginning.

The first jacket that the shop assistant wanted to show me can only be described as the weighted average of a Michelin mascot and a swimming pool for kids. The kind of jacket which may come in handy when your boat is about to capsize on the Baltic Sea, but mostly makes you look like someone who is about to engage in an inflatable sumo wrestler costume competition. I don't want to blow it out of proportion, but this was a big no-no: I only appreciate excessive volume when it comes to personal bubble space — which is not a euphemism for the jacket by the way — and gin in a cocktail.

The next thing he suggested was this jacket the experts are referring to as 'a wind-stopper'. First of all: I don't want to be too anal about this, but I have tried them, and if there is one thing these jackets can't do, it is effectively stop wind. Which is hardly a surprise, because if they really were capable of doing that, why don't they just coat aeroplanes with them? Secondly: the only jacket I would consider buying because of its ability to stop something, is the one capable of stopping people from plastering my mailbox with flyers advertising cheap pizzas and kebabs. So once again: no, that was not what I was looking for.

Once the persistent shop assistant started showing me jackets that looked like atmospheric diving suits, I knew I wasn't going to find something to my liking. On the contrary, I realised that even if I were to find a model that suited me — within my budget — there would always be that last thing that screws it. The colour.

Seriously, why is it that companies like Jack Wolfskin and The North Face spend thousands of dollars to have a team of engineers design a jacket in which you can survive anything from heavy blizzards to blazing hellfire, but fail to understand that when it comes to buying clothes — and we are talking about the variety that is to be predominantly worn outside, where others can actually *see* you — colour is a crucial criterion for purchase?

Not that you do not have a choice though: on the one hand there are what I would call the posh colour varieties, reserved for the cotton sweaters you typically see during late, chilly summer evening strolls along the Mediterranean coast: loosely dangling from the owner's shoulders. On the other hand, there are the colours that are so flashy that you could walk into a darkroom wearing them and spoil a photographer's working day. That is the equivalent of having to choose between being reincarnated as a laboratory animal in a company where cactus-shaped butt plugs are tested, or as a desert rat suffering from sand allergy.

'Relax, the colour is just a minor detail,' I hear you thinking. I couldn't agree less: let's not forget that amongst the people who are actually buying and wearing these thick fluorescent hyper-expensive jackets, you find a fair number of researchers partaking in Arctic field trips. I am pretty sure that 'flashy colourful jacket' means 'I am here' in any predator's language system — including the ones that are bigger, meaner and hungrier than you. The only logical conclusion I can draw — the power of scientific deduction at work here, or me having watched one detective series too many — is that polar bears are indeed colour blind. Q.E.D.

Recommended listening

Artist	Song title
Nirvana	Come As You Are
Squarepusher	I Wish You Could Talk
Cult of Luna	In Awe Of
Wrekmeister Harmonies	The Gathering
The Dillinger Escape Plan	Room Full of Eyes
Fuck Buttons	Okay, Let's Talk About Magic
Faith No More	Evidence
Snap	The Power
Glowsun	Inside My Head
Grauzone	Eisbär

7

Space To Slice And Stack

> *You better cut the pizza in four pieces*
> *because I'm not hungry enough to eat six.*
>
> (Yogi Berra)

You should really read this chapter if ...

- you are sick and tired of cutting pizzas the old-fashioned way.

- you're secretely afraid of more than three dimensions.

- you believe that 'the Evil Pizza' is a horror movie title.

- you're about to tile your kitchen floor in nine dimensions.

7.1 Slice matters

By far the most common situation in which I find our universe *just* not spacious enough to my liking, is when I have to clean up my apartment. I have accumulated so much stuff over the years that I no longer know where to put it. I suppose one of my problems is that I cannot really throw things away — even scrap paper containing back-of-the-envelope calculations and notes to myself are condemned to spending the rest of their lives on my kitchen table, fading away in the sunlight — which means that in my case, cleaning often amounts to organising my stuff into wobbly piles and shuffling these around. It is not so much cleaning as it is a postmodern household interpretation of the spinning plates trick.[95] I often wish I were working at a space agency, so I could

---

95   Then again, you can't clean. You can only make other stuff dirtier.

shoot my stuff into an orbit around the Earth. I bet this is what Elon Musk's stunt from February 2018 was all about: before he shot the Tesla Roadster into space, he probably filled its boot with rubbish. There is one cheaper — albeit slightly probabilistic — option as well: you can put rubbish in your socks before throwing them in the washing machine, as this guarantees half of it will disappear.

You may think that people shooting their personal belongings into space is a waste of space budget, but then I assume you've forgotten about that time when the pizza marketing wars officially reached a point of cosmic proportions — pardon the pun. On September 1st 2011, Domino's Japanese branch announced plans to open the first pizza restaurant on the moon. Just in case you think this is a figure of speech, I do mean the celestial body in orbit around our planet. Our nearest galactic neighbour. Home of the moon rabbit — according to East Asian mythology — and two golf balls, the result of a black hole-in-one gone terribly wrong, I assume.

Definition 47.

Pizza: *a special kind of fast food that was invented by a team of dieticians and Dutch toxicologists, sponsored by the World Health Organization in order to find a cure against the munchies.*

It all started in 2001, when the enemies from Pizza Hut delivered a pizza to astronauts orbiting the Earth in the International Space Station. Can you imagine the astronauts' faces if it had come with pineapple? Not *just* a pizza of course: the creation of the world's very first pizza in outer space was the product of an intense collaboration between Pizza Hut and Russian food scientists. This is a quote from the official report: 'Before final certification for consumption was given, the vacuum-sealed Pizza Hut pizza had to undergo rigorous stabilised thermal conditions to determine life span and freshness-stay.'

I am pretty sure that if you were to feed this into a Scientific Bull-shit Translation Device, it would probably spew out: 'Sorry, I lost the first 7496 pages of my Priority List for Global Problems Humanity is Facing,[96] so I decided to move on to the next page.'

Domino's first counterblow, which dates back to December 2010, was to announce their plan to pay the winner of a contest 2.5 million yen (a little under 20,000 euro) for one hour of delivering pizzas. One small job for man, a giant leap for his paycheque. Even if this task involved bringing a pizza to the caretaker of that building where they lock all the pit bulls who failed miserably at the dog training school — wearing a bacon suit, sausage shoes and a shirt made of cats — it would still be worth the effort.

And that was only the first step towards pointless pizza insanity. Some time later, Tomohide Matsunaga — a spokesman for Domino's — revealed the company's ambition to build a pizza dome on our sole natural satellite, as they anticipate there will be lots of people living on the moon at some point: astronauts working there, students distributing pizza flyers and Dutch tourists (once Elon Musk has figured out how to shoot caravans into space). The company added that they even expect to be able to offer delivery services. I didn't see the point in that at first, but I later realised that the people on the moon will be pretty high up there — and I am not just talking about the astronaut from the first chapter, celebrating his birthday with a bit of space in his cake. So yeah, pizza on the moon — in hindsight it's actually hard to believe that NASA (the National Aroma and Scent Administration) hasn't thought of this before.

In retrospect, I regret not having been part of these think-tanks: anyone who seriously considers shooting a pizza into space would probably be open to my ideas for the future as well. Ideas along the same lines of madness, if you ask me. Things like colouring books for kids with Arctic landscapes — which may not be as pointless as

---

96   For all I know, it might be on my kitchen table, fading away in the sunlight.

it seems at first sight: the jackets do require some work, remember (see the previous chapter if not)? Or how about remote-controlled touchscreen televisions and umbrellas made of sponge?

Definition 48.

Umbrella: *if there is one thing less adequate to challenge the combination of rain and wind than a giant sponge the size of a queen-size mattress, it must surely be an umbrella. More often than not, people not carrying an umbrella get less wet than the ones who do, as the latter find themselves exposing their backs to the lashing wind — trying to hold on to an item that looks like a small parachute turned inside out — hence spending more time in the downpour than the speedy non-carriers.*

The very fact that umbrellas — objects which are specifically meant to be taken outside — are designed using pieces of foldable metal strongly suggests that their inventor was born in a country where a mild breeze is considered to be a rare meteorological phenomenon.

Compared to Europe, you see surprisingly few people with an umbrella in Southeast Asia, a region known for its monsoon season. This obviously has to do with the difference between European showers and monsoon showers. In Europe they can last for hours, days or even weeks. In Asia, you basically get the same amount of water but in a shorter period — as if someone twisted that little ring on top of the modern shower head from 24 equally distributed soft rays of water to that single gush, coming straight through the middle, giving people wearing a shower cap some sort of mild waterboarding experience.

Funnily enough, once the gates of Flood City are open, Southeast Asian people tend to use whatever item they can find to shield from the rain. Grocery bags, briefcases, jackets, shorter countrymen, and one of the most common things people carry around: things you can read. Even I would consider holding one of those glossy fashion magazines over my head when it starts raining — I sometimes wonder how much sturdier you can make a piece of paper without having to label it as plastic which may contain traces of paper — but I would never use a newspaper. In sharp contrast to Southeast Asians: I have seen more than one local with the pulpy remnants of a gazette spread over his skull, the mirror image of today's news printed across his face. I suppose this is why they call it a headline.

Outer space is not the only place where you might not expect to bump into a pizza, as there are quite a few mathematical results concerning the cheesy foe of fried chips in the fierce comfort food competition. These results often have something to do with slicing: the fast food version of a problem known in maths as 'dividing a space into subspaces'. And the word 'space' is not used in vain here, as it again concerns a set with extra structure defined on it. As a matter of fact, we will work with so-called inner product spaces in this chapter, because cutting and slicing requires that you can talk about distances and angles. Going back to our earlier meta-metaphor in which we identified abstract spaces with countries on a map, we could say that inner product spaces and metric spaces are very close to each other. You can even say how close, because they both come with a ruler. There is a crucial difference though: only the former comes with a protractor too, allowing you to measure angles and speak about orthogonality. So if you ever decide to buy a holiday home, you may want to consider Inner Product Land rather than Metric Land: always handy to live in a house whose walls are perpendicular to the floor.

Mathematicians can slice spaces in any dimension m, but since we were talking about pizzas earlier we will focus our attention on the case $m = 2$. Some people might feel tempted to point out that a pizza is a three-dimensional object. Technically speaking, these people are right. It even forms the basis for a toe-curlingly bad maths joke, for which I claim no responsibility — at all. What is the volume of a pizza with radius z and height a? Easy, that would be $\pi \cdot z \cdot z \cdot a$ of course. Bad joking aside though, I have two things to say about the dimensionality of a pizza. First of all, it can't take up too much space in the third dimension: if your pizza is not flat, chances are that you are confusing it with quiche or focaccia. Secondly, I think it's safe to assume that literally no one in the history of mankind has *ever* attempted to slice a pizza along the third dimension — because I think we would all know this guy as the one who redefined the word 'simpleton'.

There, for instance, exists a result in (classical) geometry — the kind which governs our Euclidean universe, where parallel lines are monogamous and triangles stick to 180 degrees — which is called 'the Pizza Theorem'. It tells you when a specific way to slice a pizza for two is 'fair' or not (equal shares). The way people usually slice pizzas is by making cuts from crust to crust which go through the centre of the pizza. Mathematically speaking, this amounts to drawing a diameter in a circle. If you do this in such a way that the angle between two consecutive cuts is equal, this will always give a fair division into an even number of slices. By this we mean the following: if you were to number all the slices in consecutive order, then the total area of all the even-numbered slices will be the same as the total area spanned by the odd-numbered slices. For example, cutting the pizza four times gives you eight slices in total, each spanning an angle of 45 degrees. Let's label these slices in a clockwise direction. If you eat the odd-numbered slices (1, 3, 5 and 7) and I then take the rest (2, 4, 6 and 8), we will both have exactly half of the pizza.

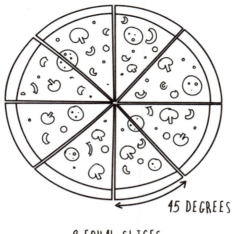

45 DEGREES

8 EQUAL SLICES

Slicing pizzas: the classical way

But you can also make cuts which meet each other in another point, somewhere off-centre, in such a way that the angles between the cuts are still equal. If two people now again take alternating slices, the pizza theorem tells you who will end up with the largest amount of pizza.

Not unlike surgery, this will crucially depend on the number of cuts you make. If you cut only once (into two slices) or twice (into four slices), it is easily seen that the result can never be fair if you do it off-centre: the person who takes the piece containing the middle of the pizza will always get the lion's share. Which is not always the best option, by the way: I'd rather have the other share if I know that the zookeeper is about to enter the room, patting a giant feline creature on the head, pointing me out as the guy who took its share.

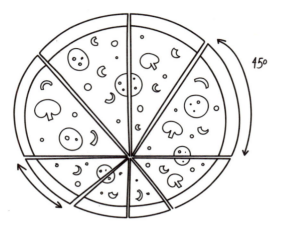

STILL 45 DEGREES: SLICES NOT EQUAL, FAIR DIVISION

Slicing pizzas off-centre

However, if you make an *even number* of cuts (so with $n = 2k$), then the resulting $2n$ slices of pizza will always lead to a fair division of the pizza (except for $n = 2$, as mentioned above). For instance, four cuts ($n = 4$) will divide the pizza into eight 'fair slices'. Strangely enough, this does not hold if you make an *odd number* of slices: three cuts lead to six slices and one person getting more than the other. Simple observation has taught me that a similar conclusion holds for the number of wheels on a vehicle when it comes to drawing people's attention: unicycles and trikes still seem to beat bicycles and cars.

Definition 49.

Pizza theorem: *this mathematical result in Euclidean geometry should not be confused with 'the Banana Chocolate Cream Pie with Salted Caramel and Rum Soaked Cherry Coulis Theorem'. This open problem says that no matter how the pie is sliced, there will never be a fair division as everybody ends up wanting more and claiming that they got the smallest slice.*

7.2 Food fight

So how about we now get rid of equal angles and crust-to-crust cutting? In other words: this time we put our knife in the centre of our pizza, and we are allowed to make any number of straight cuts from this point to the crust (formally, we may thus draw a number of radii rather than diameters in a circle). Moreover, although your mother may have taught you that food is not meant to be toyed with, we are going to turn the consumption of our pizza into a game for two people with the following rules:

- players pick a slice in alternating fashion,

- 'thou shalt covet thy neighbour's slice': players always have to choose a slice which is adjacent to the part of the pizza that was already eaten (so with the exception of the first and the last slice, players can always choose between two slices: left or right of the gap).

Now let us assume that you may start the game, thus choosing the first slice. Here is the question: can you always end up with at least half of that pizza, and how can you accomplish that?[97]

There actually exists a whole branch of mathematics devoted to the gentle art of playing games — and making enemies — most aptly named 'game theory'. In some universes, this field was introduced by a man named Jack, who was so sick of hearing that all work and no play made him a dull boy that he decided to turn the latter into the former, but these are all parallel ones: in the one where you are reading this book, game theory effectively gained its place on the list of mathematical disciplines in 1928, when John von Neumann wrote a seminal paper that became a classic in the world of mathematical economics — one of the areas where game theory is effectively applied, together with political science,

---

97   A game with cheese, is like a swordfight: you must think first, before you choose. ('Da Mystery of Cheeseboxin' by The Wu-Tang Clan)

biology and domestic discussions. You may have seen A *Beautiful Mind* with Russell Crowe, a movie about the Nobel Laureate John Nash who did revolutionary work on game theory (the 'Nash equilibrium' was named after him). A typical question game theorists try to answer is the following: given a game, such as the one described above, does there always exist a winning strategy, a set of moves which leads to guaranteed victory?

Definition 50.

Game theory: *an advanced mathematical framework centred around the following conundrum: is it possible to play a game of Risk without blowing your friendship to smithereens?*

So let's get back to our pizza game: given that you may choose the first slice, is there a winning strategy? When there is an even number of slices, the answer is clearly *yes*. Here is why: suppose we put an olive on every other slice. As there is an even number of slices, there are as many slices with an olive as there are slices without one. Hence, there are two possibilities: either the slices with olives represent at least half of the pizza, or the slices without olives do so (the areas can also be equal of course, which would be quite a coincidence as you started from arbitrary cuts, but that does not change the argument: in that case you just choose based on your preference).

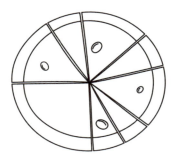

OLIVES FOR THE WIN !!

The pizza game in case of an *even* number of slices

For instance, if the slices with olives add up to more than half of the pizza, you just start with picking such a slice. Your opponent is now forced to choose one of two neighbouring slices, without olives. Regardless of his choice, you can always pick a slice containing an olive. Hence, you end up with all the olives and we assumed that this was at least half of that pizza. *Presto!*

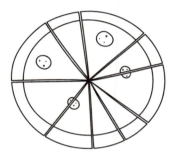

The pizza game in case of an *odd* number of slices

When there is an odd number of slices, it feels like the player starting the game can *definitely* get more than half of that pizza, right? Because an odd number is an even number plus one extra slice. Strangely enough, there are peculiar pizzas for which the second player will actually end up with at least half of that pizza, if he plays it smart and follows a prescribed strategy, regardless of how you — the first player — pick the slices. So when it comes to the pizza game, the overall conclusion is that there is *no* strategy which ensures the first player will end up with the bigger portion. Unless 'inviting gluten intolerant opponents' counts, that is.

Definition 51.

Gluten: *this synthetic food additive was conceived right after the invention of the sandwich. It is a glutinous substance (hence the name) which allowed bakers to glue individual sandwiches together, which then led to the birth of what is now known as 'bread'.*

Mathematical conclusions like the one from the previous paragraph — let me remind you that we started from a game, involving fast food — somehow make me think of fruitflies: where precisely do they come from? As it turns out, people actually wrote a research paper about this problem: 'How to eat 4/9 of a pizza', an elaborate confirmation of an earlier conjecture made by Peter Winkler. The latter correctly predicted that the first player can always eat at least 4/9, just a tad under half a pizza, but in some cases this is the best that one can do.[98] Honestly, this is one of the reasons why I fell in love with mathematics in the first place: you start with a simple question, based on an innocent observation, and before you know you it are dealing with a deep problem. Not as deep a problem as the Hodge conjecture or rescuing a Thai soccer team from a cave, but still: deep enough to have mathematicians write a paper about it.

I will not bother you with all the technical details here, but if your curiosity is as uncontrollable as a buttonless TV remote, you can always look up the final theorem on the internet. Instead, I want to give you an example of an oddly sliced pizza for which the second player indeed has a strategy which secures him more than one half. Given the fact that this means that you, the starting player, end up losing the pizza game, it may be comforting to know that it will be more than a *cold* half: his strategy does require some thinking. Note that the size of each of the slices in the picture below is not up to scale, only the numbers in the slices represent their 'true size'. But including a properly scaled illustration here would be like walking into Mr Nicholson's toilet while he is taking care of business: you'd see jack shit. So there's a bunch of tiny slices (size 1), and a few huge ones (sizes 100 and 200).

---

98   I suppose this is why you always have the option to order garlic bread on the side: it makes up for that tiny deficit.

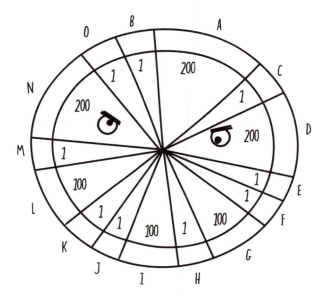

An example of an evil pizza

In the table below, I have simulated two examples of games based on this pizza: the letters refer to the chosen slices, and the numbers to the corresponding size. There are 15 slices in total, so eight for the player who starts and seven for the opponent. Moreover, I started with a maximal slice for the first player, because I assume this is something everyone would do (it feels like getting a head start). After all, playing the pizza game is a bit like drafting a basketball team: it's just weird if you don't pick the big ones first.

So let's say player 1 starts with slice A, one of only three options if he wants to start with a big slice (he obviously does not *have* to do this, I merely do this because intuitively it feels like a good start). It is immediately clear that a smart player 2 will then avoid slice C and choose B instead, because this forces player 1 to pick a slice neighbouring a big one. Either he has to take slice C (as in the left table below), or slice O (as in the right table). In both cases the game now more or less dictates itself, although the tables below are merely two of several possibilities. And when I say 'several', I actually mean $15 \times 2^{13} = 122,880$ games in total. Indeed, 15 slices to start with, followed by 13 moves in which a player has two options.

Player 1		Player 2	
A	200	B	1
C	1	D	200
E	1	F	1
G	100	H	1
I	100	J	1
K	1	L	100
M	1	N	200
O	1		
	405		504

Player 1		Player 2	
A	200	B	1
O	1	N	200
M	1	L	100
K	1	J	1
I	100	H	1
G	100	F	1
E	1	D	200
C	1		
	405		504

So according to the scenario above, player 1 will end up with 405/909, whereas the opponent gobbles down 504/909, which is indeed more than half of that evil pizza. You can actually try out a few variations for yourself (122, 878 to go) and make some tables like the ones above: you'll see that the second player can always win if he or she chooses wisely.

Solved Games: proving that a player can always win, lose or force a draw in a game is one of the main problems game theory is concerned with. It even seems that the hunt for an optimal strategy is a topic in which both mathematicians and logologists (people studying recreational linguistics) are interested. A nice example of a solvable game is Ghost, described on Wikipedia as 'a word game in which players take turns adding letters to a growing word fragment, trying not to be the one to complete a valid word. The player whose turn it is may, instead of adding another letter, challenge the other player to prove that the current fragment is actually the beginning of a word. If the challenged player can name such a word, the challenger loses the round; otherwise the challenged player loses the round. So it's not about getting rid of the slices this time, but all about adding some.

Ghost (as a two-player game) has indeed been solved, whereby the result (obviously) depends on the word list or dictionary used as a reference. I mean, if you use a Czech dictionary you don't even need to worry about adding vowels every once in a while. A few of these strategies appeared as papers in the journal *Word Ways*, a quarterly magazine on logology. For instance, in an aptly named paper 'Ghostbusting' (1987), Alan Frank showed that if one agrees to use the Official Scrabble Players' Dictionary, the only safe letter to play by the first player, is the letter H. For any other starting letter, the second player has a winning strategy: she only needs to build towards a word from a certain list (see the paper). You play F? I play J, and this will force the game towards 'fjeld' or 'fjord'. So you lose. I guess therein lies the crucial difference with algebra, which can also be seen as some sort of game with letters: at least algebra isn't boring if you know the strategy.

Just in case you thought that's where it ended: there are many variations of the original game of Ghost, and each of these comes with the quest for a winning strategy. There's Superghost, Superduperghost, XGhost, Spook and even German Ghost. The latter is based on the idea that German words can be formed quite freely by concatenation, so that every word — albeit nonsensically — is technically allowed. I suppose Ghost is like soccer: it's a game, and in the end Germany wins.

I am not sure whether Einstein had phrases like 'I had a good day at the office' in mind when he suggested that everything is relative, but I think they do qualify: their meaning ranges from 'my boss didn't call me a complete nitwit today' to 'I was dealt pocket aces, went all in and won myself a 10,000 dollar poker tournament during the lunch break'. You'd think that halfway across that spectrum, you'd see 'we finally proved Winkler's pizza conjecture', and that this would be the end of the story. But that's not how the mathematical mind operates. Not unlike power hungry emperors, mathematicians are known to continuously push the boundaries, and so they are always trying to generalise a problem once they have found a solution. They solved Ghost? No problem, they invent Superghost. They understood Pythagoras' theorem (for $n = 2$)? No problem, they try to tackle arbitrary natural numbers n (Fermat's last theorem). I guess mathematicians use the same rhetoric as drunk aggressors who want to pick a street fight: 'if you want trouble, you can get trouble.' The pizza problem forms no exception to this rule, as the following questions are genuinely considered by maths researchers:

- Can we still come up with a strategy for the first player if the pizza is allowed to have slices of negative size, but in such a way that the total size of the pizza is still positive? I have absolutely no idea what they mean by negative slices — apart from tofu, I cannot even imagine zero size food — but it might become a big thing with weight watchers.

- Similar to the original pizza game, but a player is only allowed to take the next piece when he or she has finished eating the piece in hand. This problem is known as 'the pizza race problem'. Not to be confused with pineapple, 'the pizza disgrace problem'.

- Chances are you never thought about this before, but mathematically speaking a pizza is nothing but 'a simple cycle graph'. It's just that this name doesn't combine well with Margherita, because I am pretty sure it would have stuck otherwise. Representing each slice as a dot — or, in maths lingo, a vertex — you can see why: each vertex is connected by an edge with its two neighbours, hence creating a closed chain. In doing so, you will end up with the cycle graph C_6 of order 6. This may all look reductionistic, but so is a pizza Margherita.

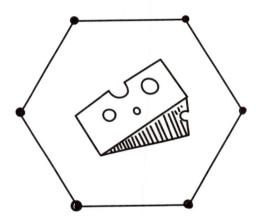

EVEN PIZZA ABSTRACTIONS LOOK
BETTER WITH SOME CHEESE...

So one way to generalise the pizza problem, is to switch from cycle graphs to more complicated graphs, and to investigate how one can play the pizza game on these graphs. This may seem far from the reality of your kitchen, but there is a message to keep in mind here: next time your baking project turns out to be an epic flour and eggs fail, you can always claim you were trying to make a hexagonal truncated trapezohedron graph with mozzarella and anchovies.

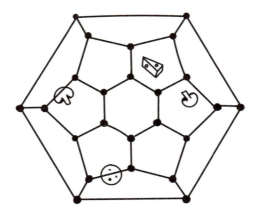

Your next epic flour and eggs fail

Definition 52.

Graph theory: *a branch of mathematics in which graphs are studied. A paper from 1736 by Leonhard Euler entitled 'Seven bridges of Königsberg' is often regarded as the first paper in graph theory. The city of Königsberg, now Kaliningrad, was set on both sides of the Pregel River and included two large islands which were connected to each other and the mainland portions of the city by seven bridges. Some people wanted to know whether it was possible to walk through the city in such a way that each of those bridges was crossed once, and only once. Euler himself proved that this is actually impossible, but the main lesson I learnt from this is that even people suffering from OCD can come up with interesting maths problems.*

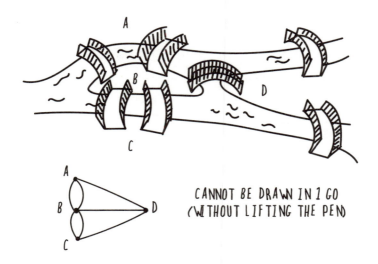

CANNOT BE DRAWN IN 1 GO
(WITHOUT LIFTING THE PEN)

From bridges to graph theory

7.3 Catering for the lazy

It is not all games and well-thought-out strategies though: sometimes that pizza comes out of the oven, and we want to dig in as quickly as possible. All of us, at the same time. This is exactly what *the lazy caterer sequence* is all about: it describes the maximal number of pizza pieces (not necessarily slices) that can be obtained using a certain number of straight cuts. Cutting a pizza once gives you two pieces — unless you're Zorro. Cutting it twice already shows that the number of pieces you end up with crucially depends on where you put that wheel slicer blade: if the lines along which you cut meet each other, you get four pieces. However, if the lines are parallel for instance, you only get three pieces. Well, it can still be more if you're a bit careless with the knife, but in this case the fingers don't actually count.

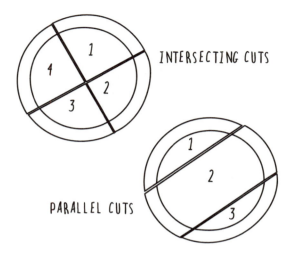

So here's the question: if you make n straight cuts, what is the maximal number of pieces you can end up with? If all the cuts meet each other in the same point — for instance in the centre, the classical way of slicing a pizza — you get $2n$ slices, but can you do better than that? Solving a mathematical problem depending on a parameter $n \in \mathbb{N}$ is like trying to win a rugby match: you can't do it if you don't tackle a few numbers. We had $n = 1$ and $n = 2$, leading to a maximum of two and four pieces respectively, so let's try the next case. For $n = 3$, it's not too difficult to see that you can end up with at most seven pieces of pizza. The idea is that you start from two intersecting cuts, which already gives you four pieces, and that the third cut will either go through the point of intersection (giving you six) or meet both previous cuts exactly once (giving you seven).

As for the general case, the clue is in the last sentence of the previous paragraph: each new cut should intersect all of the previous cuts, but not cross any of the points where these previous cuts intersect. This means that the last cut (number n) intersects all of the other $(n - 1)$ lines, and thus adds another n pieces to what we already had. Think of the case $n = 3$: the third cut intersects two lines and adds three pieces to the four we already had. Suppose we call $P(n)$ the maximum number of pieces after n cuts, then this paragraph can be summarised in a compact formula:

$$P(n) = n + P(n-1).$$

This is a nice example of a so-called *recurrence relation* — not to be confused with going back to your ex — which is the mathematical equivalent of a domino show: once you topple the first tile, all the rest will follow. And in order to predict what the nth tile will do, you only need to know what the previous one has up its sleeve. That is what the formula says: the value $P(n)$ is known as soon as you know $P(n-1)$, it suffices to add n.

> *'How much is it then?'*

Well, all you have to do now is to peel away the layers: obviously you need to know $P(n-1)$ too, but according to that very same recurrence relation we have that $P(n-1) = P(n-2) + (n-1)$. This goes on for a while, until you reach the first tile,[99], as this is the one that started the (pizza) chain reaction. In this case, it all starts with the value $P(0)$, which is equal to one: if you don't cut it, the complete pizza counts as a single piece. Putting everything together, we thus get that

$$
\begin{aligned}
P(n) &= n + P(n-1) \\
&= n + (n-1) + P(n-2) \\
&= n + (n-1) + (n-2) + (n-3) + \ldots + 2 + 1 + P(0) \\
&= (n + (n-1) + (n-2) + (n-3) + \ldots + 2 + 1) + 1
\end{aligned}
$$

> *'But how much is it?!'*

---

99   Intensively stared at by the guy from a previous chapter, still desperately trying to prove the non-existence of unturned stones.

Well, it seems that we are now left with the following problem: which number do you get when you add up the first n integers (this is the number between brackets on the last line of the formula)? Legend has it that Carl Friedrich Gauss came up with a clever trick to do this. According to the story, his teacher got mad in class one day and made everyone add up the first 100 numbers. He thought this would keep them busy for a while, but it took the young Gauss less than a minute to come up with the number 5,050, which is indeed the correct answer. Not only that, he even came up with a formula that works for any number $n \in \mathbb{N}$. I guess this proves that alternative punishments do work. It definitely opens up new perspectives for judges: 'I hereby condemn you to ten years in solitary confinement. Or you can prove the Goldbach conjecture.'[100]

The unruly Gauss came up with the following rule: in order to add up the first n integers, it suffices to look at the following problem:

$$1 + 2 + 3 + ... + (n - 1) + n$$

$$n + (n - 1) + (n - 2) + ... + 2 + 1$$

The numbers in a column add up to $(n + 1)$, and you have exactly n of those. However, you added them all *twice*, so it suffices to divide by 2 to arrive at the final result:

$$1 + 2 + 3 + ... + n = \frac{1}{2} n(n+1)$$

Getting back to the lazy caterer sequence, it seems like we are now finally able to solve the original problem. Given a pizza, you can use n straight cuts to end up with a maximum number of pieces given by the number

$$P(n) = 1 + \frac{1}{2} n(n+1)$$

---

100   Then again, Andrew Wiles will probably say that they are pretty much equivalent.

For instance, cutting three times ($n = 3$) gives seven pieces, since

$$P(3) = 1 + \frac{1}{2}(3 \times 4) = 7 \, .$$

So far, we have been talking about slicing pizzas in a variety of mathematical problems, but it seems we are overlooking the biggest issue mathematicians have with pizzas: preparing them. The thing is that pizzas have to be baked in an oven at 225 degrees (or, as mathematicians would say, $\frac{5\pi}{4}$ radians), and this is difficult for them: when it comes to round things, they are used to shamelessly adding a multiple of 360 degrees.

7.4 Lost in translation

Chances are that you took the evil pizza game from a few pages ago a bit too seriously, which means that you may now be staring at 37 empty pizza boxes.[101] No worries, we can put them all to good use. Because not unlike cats, mathematicians are extremely fascinated by boxes. Even something as simple and innocuous as stacking boxes, or tiling floors (the two-dimensional equivalent), has the potential to boggle mathematicians — especially in higher dimensions. So buckle up, it seems the time has come for a trip into hyperspace.

Theorem 4. *Do not hire people to tile your kitchen floor if you happen to live in a space of more than nine dimensions: the result may be rather erratic.*

---

101 Please don't tell me you went for the full 122, 880 verifications.

Proof: Let us start from an infinitely large kitchen.[102] If this is too difficult for you to imagine, feed your brain the following program:

```
for i = '1' to 'Okay, I get it now - it goes on forever' do
    Kitchen := Kitchen + one cubic metre;
        i := i + 1;
    end.
```

Well, I was assuming you had trouble with the boundlessness here. If it is the kitchen that troubles you: it is that room where all the people gather when you throw a house party. Now suppose that the floor of this infinite kitchen needs to be tiled, with an infinite collection of equal-sized squares. No matter how hard you try, when tiling an infinitely large kitchen, you will always end up with a floor with the property that every square has at least two complete edges in common with its neighbours. Just in case you've *really* got into topology in the meantime: yes, it has to be a square. Circles and triangles won't suffice this time.

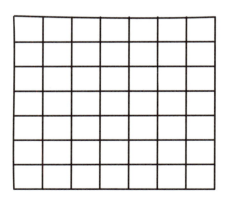

The 'normal' chessboard configuration

---

102 As far as dramatic opening lines go, this one probably still doesn't beat Genesis — or does it?

Indeed, either the kitchen floor ends up like a boundless chessboard (which means that every square shares complete edges with each of the four neighbouring tiles), or it ends up like a kind of deranged chessboard in which every row is translated with respect to neighbouring rows, which means that every square still shares complete edges with two of its neighbours. Note that the situation is fundamentally different for finite kitchen floors, in which the latter configuration can never be accomplished, unless you start breaking tiles to fill up the gaps near the edges. But in the problem we are dealing with now, our floor is *infinitely* large, so there are no edges — only endless frustration when it needs to be cleaned.

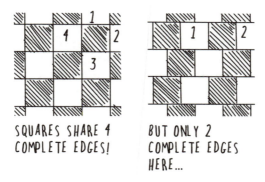

SQUARES SHARE 4
COMPLETE EDGES!

BUT ONLY 2
COMPLETE EDGES
HERE...

The upshot is the following: in the case of two dimensions (referring to your standard kitchen floor) these are the *only* two possibilities. Any other arrangement of tiles will have rectangular 'holes' which cannot be filled up, unless you start breaking or cutting tiles — even when it is infinitely large.

Next, in that vast kitchen — with its infinitely large drawer under the kitchen sink, housing an infinite collection of 5 cent plastic bags which you keep forgetting when you go shopping — you have an infinitely large fridge and a ditto freezer. Suppose now that this freezer needs to be filled with an inexhaustible number of equal-sized boxes.

In other words, we are going to raise the stackables: from square tiles (two-dimensional objects) to cubical boxes (three-dimensional objects).

Again, you will always end up with a configuration in which every box has at least two complete squares in common with its neighbours — even though the final configuration of boxes can be good, bad or downright ugly (so there will be three options this time). To see this, you can think of the present situation as an infinite collection of infinitely large thick chessboards, all stacked on top of each other. This means that the equivalent of the perfect chessboard arrangement from before now becomes a perfect, infinitely tall tower. The result then looks exactly like what you see when you open a box of sugarcubes: perfectly stacked, in such a way that each cube shares complete squares with each of its six neighbours (up, down, left, right, in front and behind). This is *good:* an infinite stack of untampered, perfectly aligned chessboards.

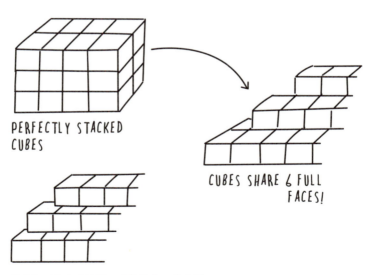

PERFECTLY STACKED
CUBES

CUBES SHARE 6 FULL
FACES!

CUBES NO LONGER SHARE COMPLETE
FACES FROM ABOVE & BELOW

But you can start meddling around in two steps now. First of all, you can keep the untampered chessboards, but you can stack them in such a way that they no longer form perfectly aligned towers. This is the *bad* case: you shift complete layers, causing the boxes to no longer share complete edges with the neighbours from above and below. So there are only four shared faces left now — just like the chessboard, but for boxes instead of squares.

But you can also give up the untampered chessboards, and shift the rows of boxes in a fixed layer. This is the *ugly* configuration: a tampered tower of tampered chessboards. At worst, each of the boxes still has complete squares in common with just two of its direct neighbours. Further derangements of the boxes are not allowed, as they will again create holes which cannot be filled without cutting some of the boxes.

Proceed with caution now, as this is the point where a stretched mind might end up being a strained one. Because you can repeat this idea in any dimension, by stacking hypercubes.

< The sound of someone begging for a pardon >

Hypercubes are the higher-dimensional version of squares (in two dimensions) and cubes (in three dimensions). Before you start kicking your own butt because you don't know how a hypercube in, say, 17 dimensions looks like, let me assure you: neither do I, and nor did Einstein.[103] Then again, I would not recommend kicking your own butt under any circumstances, unless you got bored because you figured out how to sleep on n ears with $n = 2$ and lick your elbows.

---

103   I noticed that dropping Einstein's name in an argument endowes it with a certain sense of authority. The opposite can be achieved by mentioning Donald Trump in the same sentence.

As mentioned earlier, when I talked about Euclidean geometry, this is not your fault. Our brains got wired in such a way that we simply *cannot* do this, because we were born in a three-dimensional universe. Please ignore the murmuring in the background, it's just the string theorists. You cannot even learn how to do it; we simply evolved into beings who cannot handle more than three dimensions. Which just goes to show that natural selection is somewhat like a friend with benefits: it comes with sex, but it does have its downsides.

But despite the fact that we are no longer able to *visualise* hypercubes, mathematicians still have a framework to describe and work with them. You see, the current credo governing our social lives since the advent of Facebook and Instagram and the like — pictures or it did not happen — does not apply to mathematicians: they do not need pictures, as they have alternative ways to visualise things, in their mind's eye.

One way to work with hypercubes, for instance, is to resort to coordinates. To be more precise, I should say *Cartesian* coordinates here, as there are different kinds of coordinate systems. We already met the polar coordinate system in an earlier chapter, but one can also use cylindrical or spherical coordinates, homogeneous or Plücker coordinates and a few other exotic species reserved for special occasions — like fondue forks in a cutlery rack. As for now, we will stick to the coordinate system introduced by René Descartes, the French philosopher, mathematician and scientist (1596-1650) — there are less appealing titles to put on a business card — known for his '*cogito ergo sum*'.

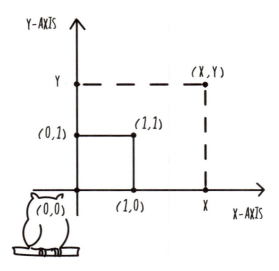

Coordinates defining the square in two dimensions

Let us once again focus on the square in two dimensions first. If we draw *four* points in the plane (often referred to as 'vertices') with coordinates of the form

$$(0, 0),(1, 0),(0, 1),(1, 1)$$

and then connect those vertices whose coordinates differ in one position only, we end up with a square. Note that not all the vertices are connected. For instance, the vertices (0, 0) and (1, 1) are *not* connected: this would give you a *diagonal*, rather than an edge. Each vertex belongs to exactly two edges, which are geometrical objects in one dimension (parts of a line). The upshot is that you can obviously still *draw* a square, but you can also 'see' it as a collection of vertices and certain rules which tell you how to connect these.

Keeping this guiding philosophy in mind, let us consider the cube in three dimensions again. This time, we start from the coordinates for the following *eight* vertices in space:

$(0, 0, 0),(1, 0, 0),(0, 1, 0),(0, 0, 1),(1, 1, 0),(1, 0, 1),(0, 1, 1),(1, 1, 1)$

If you now again connect all the vertices whose coordinates differ in one position only, you'd end up with a cube. For instance, the point $(1, 0, 1)$ is connected to $(0, 0, 1)$, to $(1, 1, 1)$ and to $(1, 0, 0)$, but not to $(0, 1, 1)$. This time, each point belongs to exactly three faces — a fancy word for the six squares making up the cube — which are objects in two dimensions (squares). So in a sense, we see that the cube (our 'new' object) is obtained by gluing six squares (the 'old' object) together: to do so, we merely need to know that three 'old objects' have to meet in a vertex.

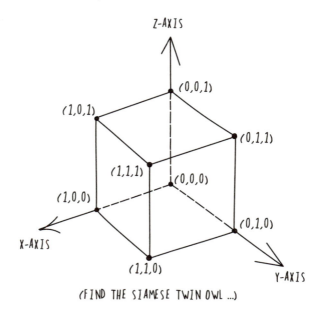

(FIND THE SIAMESE TWIN OWL ...)

Coordinates defining the cube in three dimensions

And here is where the true power of mathematics comes into play: if you look closely at the two examples above — the ones that you can still visualise using a drawing — you may see a pattern arising. And patterns happen to have the same effect on mathematicians as late night text messages on jealous partners: they ignite curiosity, spark suspicion and inevitably lead to the firm belief that something is going on. In this case, unveiling and generalising these patterns leads to abstract objects which we call *hypercubes*.[104] So to go from squares — over cubes — to hypercubes, we need a collection of points, together with certain connections between them. We will even be able to say how our hypercube (the 'new' object) can be obtained by glueing cubes (the 'old' object) together.

- First of all, we need a bunch of vertices. We went from four to eight vertices when going from two to three dimensions, so not surprisingly we will need 16 vertices in four dimensions. The question then becomes: what do they look like? Again inspired by the previous examples, the power of analogy tells us that in order to write them down, we need some sort of code. Each point will now be identified with coordinates of the form (x_1, x_2, x_3, x_4), where each of these numbers is either 0 or 1. This gives you two choices on each of the four positions, so $16 = 2 \times 2 \times 2 \times 2$ points in total. They are given by

$$(0,0,0,0), (1,0,0,0), (0,1,0,0), (0,0,1,0)$$

$$(0,0,0,1), (1,1,0,0), (1,0,1,0), (1,0,0,1)$$

$$(0,1,1,0), (0,1,0,1), (0,0,1,1), (1,1,1,0)$$

$$(1,1,0,1), (1,0,1,1), (0,1,1,1), (1,1,1,1)$$

---

104 This technique of generalising an analogy is the perfect mathematical embodiment of an Asian philosophy: 'same same, but different'.

So we have absolutely no clue what a hypercube will look like, but it will somehow involve these points — comfortably sitting somewhere in four dimensions.

- Next, you need to turn this collection of points into an object, just like we needed to make certain connections between four points (not all of them) to arrive at a square. Once again — the power of the analogy — you will have to connect two vertices if their codes differ in one position only. For instance, the vertex with coordinates $(0, 1, 0, 1)$ will have to be connected to the vertices $(1, 1, 0, 1)$, $(0, 0, 0, 1)$, $(0, 1, 1, 1)$ and $(0, 1, 0, 0)$. If you think about it, you always get as many edges from a vertex as the number of dimensions in which you are working. On the other hand, the vertices $(0, 1, 0, 1)$ and $(1, 0, 1, 0)$ will not be connected by a line, as their codes differ in four positions. These are like the 'diametrically opposed' vertices in a square or a cube.

- It's not that easy to 'see', but each vertex will now belong to exactly four 'hyperfaces', which are objects in three dimensions (you may have guessed: these are cubes, which then 'live' in one dimension lower). This is the four-dimensional version of saying that every vertex of a cube (in three dimensions) belongs to exacly three faces. Still put differently: you can make a hypercube from eight ordinary cubes, gluing them together in such a way that four cubes meet in a single vertex. Now before you book yourself a one-way ticket to Frustrationville: this is impossible. And this has nothing to do with your dexterity, it's just that you will need another dimension.

You may now grab your bucket list — taped to the hyperfridge, I assume — and tick the appropriate hyperbox, as you just 'made' your first hypercube in four dimensions. Note that this particular object, which is also called 'a tesseract', *does* have a graphical representation.

'What? You said that was impossible.'

Well, it can still be done if you use perspective, which is a miracle in itself. I mean, if you think about it, even being able to draw a cube is an ingenious feature: paper tends to be two-dimensional, so already there it seems like you are one dimension short. Look at the illustration below: the first picture shows a cube in perspective. It seems as if someone pulled at the 'front' square, dragging it along the third dimension until its resting place was reached (the square 'at the back'). In the same vein, the picture in the middle represents a tesseract in perspective: the 'front' cube was dragged along the fourth dimension until its final position was reached (the cube 'at the back'). The third picture merely represents your fridge in perspective.

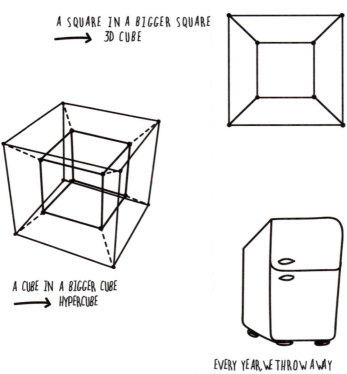

A SQUARE IN A BIGGER SQUARE
⟶ 3D CUBE

A CUBE IN A BIGGER CUBE
⟶ HYPERCUBE

EVERY YEAR, WE THROW AWAY
AROUND 1.3 BILLION TONNES
OF FOOD...

PERSPECTIVE AT WORK

Although it becomes tiresome to write this all down in (for instance) 37 dimensions, it can still be done. This will require 2^{37} vertices (which is a lot[105]) and each of those needs to be connected to 37 neighbours (the vertices differing in one position only), but the main idea remains the same. You can no longer draw it, but you can still describe it: a hypercube in 37 dimensions is nothing but a list of points, and an algorithm which tells you which dots are to be connected. This basically means that imagining hypercubes is easier than stuffing boiled eggs into your mouth, as being able to do the latter with four does not guarantuee success when increasing the numbers.

The upshot is the following: once you have figured out how to define hypercubes in 37 dimensions, you can start asking yourself whether properties that hold for 'ordinary' squares and cubes (in low dimensions) also hold for these new objects in higher dimensions, as this is one way to get a grip on their characteristics.[106] This is actually another typical feature of mathematicians: they always try to generalise — see, I did it again — and whenever that turns out to be impossible, they at least want to understand why it doesn't work.

Although hypercubes share a lot of properties with ordinary squares and cubes, like inducing four-letter words when you hit them with your knee, these infinitely large 37-dimensional kitchen floors happen to satisfy a rather peculiar property when it comes to floor tiling. In order to understand this properly, we need to go back to the infinite chessboards and the infinite towers of chessboards from a few pages ago. We observed that in those cases, you can start meddling with the boards or the stacks, but a certain amount of order was always guaranteed: each square (or cube) always had at least two sides (or faces) in common with

---

105   To be precise: 137,438,953,472 vertices. If you draw one vertex per second, this
          would take you a little more than 4,358 years.
106   Another option is to attempt a proof by LSD, but this method is still not
          widely accepted.

its neighbours. So if the power of analogy is as predictable as the end of the *Titanic* movie, you'd expect something similar to hold in more dimensions. Granted, there will be more dimensions to do the meddling in — you can push around hyperstacks of hyper-towers of hyperboards — but in the end you'd expect that there is a minimal degree of order: it is normal to expect these hypercubes to share at least two hyperfaces with their neighbours.

<*dramatic drumroll*>

And yet, things are different in higher dimensions. For instance, when tiling that 37-dimensional kitchen floor with 37-dimensional 'tiles' of equal size (hypercubes in 37 dimensions), you can do that in such a way that there is no order — at all. There are surprising configurations whereby none of the hypercubes has a complete 'side' in common with its neighbours. 'What? And there is no breaking or cutting required in 37 dimensions?' Nope, in sharp contrast to the situation in lower dimensions, it can by done by smartly organising the hypercubes. So there exists an abstract sense of freedom in 37 dimensions, envied by squares and cubes. Like a slide puzzle gone completely berserk — lost in translation.

And it gets even stranger: it turns out that the kitchen tiling property, which holds for squares and cubes, holds in dimensions four, five and six too. By which I mean that neighbouring 'tiles' will always have at least two complete 'sides' in common with their neighbours. It does *not* hold in any dimension bigger than or equal to eight though, where the same surprising property holds as in 37 dimensions: it can be tiled, but don't expect it to be neatly organised, it may very well be that the minimal degree of order is completely lost.

But there always has to be that one exception, right? That one person in your language class who says, 'After me. Ramen ga suki desu ka?' when the Japanese teacher says, 'Repeat after me: ramen ga suki desu ka?' That odd friend of yours who feels the urge to point out that it should be 'See you again today' when you are saying goodbye after midnight. The goofy colleague who genuinely believes that he just got a marriage proposal in his inbox from a Lithuanian fashion model, whose contact details can be obtained after registration on a raunchy website.

When it comes to tiling hyperkitchens — a problem which appears in the literature as 'Keller's conjecture' — we are also left with exactly one case. Of all possible dimensions: just one. The case of seven dimensions: that magical number, also known from guest appearances in both the Bible and that other fairy tale starring an apple which was not meant to be eaten, a life-altering kiss and a bunch of stubborn followers.[107]

'But who cares?'

In contrast to those midges hitting your eyeball when you go cycling along the riverbank on a summer evening: you should have seen the answer to this question coming. Mathematicians of course. Never content with partial answers, always looking for problems which may or may not be in dire need of a solution, unable to rest before everything is completely comprehended, categorised and classified. And even if they do succeed in finding a solution, interesting mathematical problems are like street riots in black American neighbourhoods: tackling one sometimes causes a more serious one to surface. So that merely brings the mathematician back to square one. And cube one. Or hypercube one. You get the picture. Well, at least the coordinates.   Q.E.D.

---

107   Homework: find further isomorphisms between the Bible and Snow White.

Recommended listening

Artist	Song title
Vanilla Ice	Ice, Ice Baby
Fatima Al Qadira	Fragmentation
Wu-Tang Clan	Da Mystery of Chessboxin'
Kyuss	Size Queen
Madensuyu	Share a Lot
System of a Down	Chop Suey!
AFX	Boxing Day
Tengil	A Box
Public Enemy	Don't Believe The Hype
Orbital	The Box

8

Too Much Space

Questioning anything and everything,
to me, is punk rock.

(Henry Rollins)

You should really read this chapter if ...

- you're curious about the role drunk wombats can play in mathematics.

- you have always wanted a fancy scientific calculator with even more buttons to press.

- you want to be able to tell the difference between angry text messages and big numbers.

- having to cook for your partner makes you nervous, because you're not really sure whether he or she will approve of your decision.

8.1 Having a ball

In the previous chapter, we studied hypercubes — the somewhat more exotic cousins of everyday squares and cubes. In this chapter, we will round off our trip into space and consider the generalisations of discs (in two dimensions) and balls (in three dimensions) to higher dimensions, known in some circles as 'hyperballs'. Once again, we'll see that objects in higher dimensions can behave rather bizarrely — and so would you, as you will soon realise. But before doing so, let us get back to the Japanese and their unique skills once more.

Theorem 5. *No need to go to higher dimensions to be fascinated by a ball.*

Proof: One of the most impressive things I have ever seen was a handmade ball, created by a waiter in a Japanese highball bar. Just in case you are not familiar with the concept of a highball: it is not a medical condition, but a cocktail made with soda water, whisky and ice. Sounds fairly simple, right?

But as with most things in the Land of the Rising Ball of Fire, nothing is as simple as it looks: even the most mundane of movements are mastered to perfection,[108] often in accordance with the underlying Zen philosophy. And since Japanese whisky is such a superior product, it should not be served with a simple ice cube, as this would stand in sharp contrast to the liquor. So when you order a cocktail in a Japanese highball bar, you get a drink with a rather special ice cube. As a matter of fact, you get a spherical cube — how does that score on the oxymoron scale? One that is so brilliant that it makes diamonds look like a girl's second best friend.

'Second best?'

Yes, second. For have you ever looked at an ice cube? I assume you have *seen* them before, but have you properly *looked* at them? 'Why should I?' I hear you thinking. 'The average ice cube is nothing but frozen water.' That's right. Conventionally used to cool beverages and easily produced at home using one of those ice cube trays which even come in a variety of shapes: cubical, cylindrical, or in the shape of a penis for that obvious pun-intended cock-tail serving.

---

108 Well, it's probably more correct to say 'perfection minus epsilon', because of that thing called wabi-sabi: beauty is imperfect and incomplete. I once wanted to buy a book about wabi-sabi, but I decided against it when I noticed that it was frayed at the edges and that there was a page missing.

But home-made ice cubes are not that pretty: they often bear scratches along the sides, they are slightly clouded in the centre and they are marked by a flight of tiny bubbles which didn't manage to escape once the water started to crystallise.

Japanese highballs, on the other hand, are served with an ice cube whose virginity is second to nun (*sic*). And the coolest part of those highballs is not even the ice ball itself, but the fact that it is hand-carved by the cocktail maker. *In situ*. Donning white gloves and using a kitchen utensil which can only be described as a cross between a machete and a samurai knife.

It all starts with an icy rectangular cuboid from the freezer. A piece of ice which is so transparent that the first time I saw it, I genuinely thought it was fake. Like in that television prank where two people pretend to carry a glass pane across the street. No scratches, no fingerprints, an unclouded center. I have no idea where they make this kind of ice, let alone how. But then again: I am a pure mathematician, which means that the sheer confirmation of existence already has the potential to bring me pleasure.[109] And from this solidified block of purity, the Japanese bartender then carves that one shape that both rules and rolls them all: a ball. From scratch, not bearing one. Dexterously wielding the sort of knife that could hurt your eyes if you look at it too sharply, producing for you the perfect period to punctuate your pleasure potion: a Japanese highball.   Q.E.D.

---

109 Some of the most important theorems in mathematics are 'existential results': they do not tell you *how* to construct this or that, they just tell you some concept at least exists. Always reassuring to know that you are not studying an intellectual ghost.

Definition 53.

Freezer: *often connected to a fridge as an even colder compartment of the installation (although there are stand-alone varieties in different shapes as well), freezers function like frozen food recycling points: people tend to collect all kinds of plastic boxes containing soup and leftover meals in them, before throwing them away. This is done as soon as they've forgotten what is inside the box, or when it was produced and put into the freezer in the first place.*

The cuboid in the theorem is technically speaking a bit of an over-kill: you might as well start from a cube, and carve from that the 'maximal ball'. By that I mean the following: there is a unique ball which sits inside the cube, touching the six sides of that cube. The smallest box around that soccer ball you bought for your nephew, so to speak — because that's just easier than wrapping it in paper. The three-dimensional version of the smallest square plate on which you can serve a perfectly round pizza.

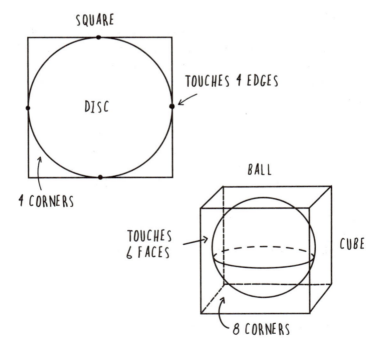

SQUARE

TOUCHES 4 EDGES

DISC

4 CORNERS

BALL

TOUCHES
6 FACES

CUBE

8 CORNERS

What we are going to do now, is to consider higher-dimensional analogues of this procedure. We have already met the material we are going to use as a wrapping (the hypercubes), but what exactly is it that we put inside them? In other words: how do we generalise the pizza and the soccer ball? Let me recall that in order to generalise you first need to see a pattern. So what do circles and spheres have in common?

<The universal whistling from someone waiting for an answer>

They are both 'collections of points' at a fixed distance (the radius) from a chosen point (the origin). 'Wait a minute, does that mean you can only define them in metric spaces? Because you are using the distance between points in your description.' Yes indeed, well done. Like a needle on a funk record: you are really getting into the groove here! As we already know that spaces of higher dimensions do 'exist' — we even know how to 'visualise' them, sort of — I merely have to give you the distance function, and that should do the job. As you may have guessed, it's nothing but the Euclidean distance. The equivalent of a circle and a ball for a *different* metric exists as well, even in dimensions higher than three, but that brings you straight into the realm of non-Euclidean hyperspaces — feel free to boldly go where no non-mathematician has gone before.

When it comes to 'normal' circles and spheres, you can think of it the following way: both objects can be seen as the collection of points traced out by a wombat attached to a rope, when you start swinging it around.[110] In two dimensions, it suffices to just spin around your own axis: that will give you a circle. In three dimensions, it requires more work, as you literally have infinitely many directions to swing that rope — lots of circles on a sphere — but in the end, you *will* get a sphere. The radius is the length of the rope, the origin is your body and the indignation of the people seeing you do it guaranteed.

---

110 Now you know where the term 'swing theorists' comes from (see chapter 1).

The sphere as the path traced out by a wombat

First of all, go to that hyperspace of your favourite dimension —
well, I am making it sound like you've got an option here, but we
all know that the only reasonable answer is 42 — catch a hyper-
wombat and attach it to a piece of hyperrope (length one). If you
don't know where to find hyperwombats, just follow the hypercube poo trail.[111] Next, you will have to get hyperactive and start
swinging that rope around. This may all take a while, as there are
plenty of dimensions to cover, but you *will* end up with a hypersphere. In other words: a collection of points at a fixed distance
(equal to one, if you did not ignore my instructions) from a fixed
point (your fatigued body). The hypersphere is the boundary, and
together with all the points closer to you (the inside) you get a
hyperball.

---

111   For the topologists reading this: never mind.

Everyday isomorphisms: *sphere is to ball like skin is to body.*

Having a ball, are we? So far, we have merely generalised the situation in low dimensions to end up with a collection of mathematical objects in higher dimensions — the ones where we can no longer rely on visualisations. But I will now do some innocent computations, and this will then lead to a rather disconcerting conclusion. I will essentially compare two numbers in what follows: a number that expresses 'the size' of a hypercube whose edges have all length one, and another number that expresses 'the size' of the maximal hyperball that can be put inside that hyperbox. But let us first consider some examples, to get an idea of what we are attempting here:

- In two dimensions, the 'size' of a square (or disc) is nothing but its area. So, for the square this gives us length times length: $C_2 = 1 \times 1 = 1$. Note that I gave the area a name for future reference: C_2. The 'C' refers to 'cube' here, and the '2' to the dimension we are working in. Technically speaking, this value for C_2 is not just a number: it is an amount of 'square metres'. Similarly, for the area of the disc we have to apply a formula which was surely rammed down your throat by a teacher wearing corduroy trouwers, occasionally lifting both heels while clasping his hands behind his back: $B_2 = \pi \times (\text{radius})^2$. The letter 'B' stands for 'ball' here, the disc merely being a special case (in two dimensions). In order to calculate this number, we need to know the answer to the following question: what is the radius of the largest disc we can put in the square? The diameter is equal to 1, so the radius is half of that, which means that the area of the disc is given by

$$B_2 = \pi \left(\frac{1}{2}\right)^2 = \frac{\pi}{4}$$

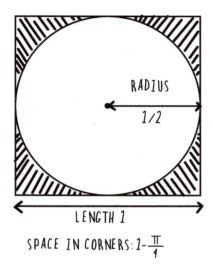

RADIUS

1/2

LENGTH 1

SPACE IN CORNERS: $1 - \dfrac{\pi}{4}$

The biggest pizza on a square plate

The number B_2 is obviously less than 1 — the area of the square — since there is some space left in the four corners.

- Let's do the same trick in three dimensions now, putting a soccer ball into a box. This time, the 'size' of the box (and the ball) is called the volume rather than the area. For the volume of the box, this gives us $C_3 = 1 \times 1 \times 1 = 1$, where I have used a similar name as above (only the dimension has changed here). As for the volume of a ball, you may remember the formula which says that $B_3 = \frac{4}{3} \times \pi \times (\text{radius})^3$. So in order to calculate this volume, we need to know the radius of the ball. As our ball touches the sides of the cube, it is again simply half of the length of the side of that box. In other words, we have that

$$B_3 = \frac{4}{3}\pi \left(\frac{1}{2}\right)^3 = \frac{\pi}{6}$$

Once again, we find that B_3 is less than 1 (the volume of the box), but that's obvious. After all, there is some space left in the eight corners of the cube.

In the previous chapter we first introduced new objects (hypercubes), and then we started asking ourselves questions that were easy to answer in the case of squares and cubes. In this chapter we introduced hyperballs, so it feels natural to start asking questions that can easily be solved for discs and balls. In particular, we are interested in finding out what the size of these things becomes when the dimension increases, and whether you will still have space left in the corners when you put them into a hyperbox. It's tempting to say 'Of course you will, that's what happens in two and three dimensions!', but one of the things you should have learnt from the previous chapter is that higher-dimensional objects are like kids: once they start getting bigger, you can't always predict their behaviour.

In other words, we are interested in knowing the values for C_n and B_n for arbitrary dimensions n. It will again be a number representing a 'size', but not unlike 'the day after tomorrow' there is no matching word for it in English (although I do think we should all start using 'overmorrow'). The reason for this is pretty simple: we survive our days working with lengths, areas and volumes only — that's the Euclidean heritage — so we rarely need 'hypersizes'. That could change once we start flying hyperplanes around the planet; I am pretty sure that even four-dimensional overhead compartments will have their limitations.

For lack of a better word, mathematicians usually refer to the 'size' of a 'hyper object' — things like hypercubes, hyperballs and other exotic bodies that can't be confined to three dimensions — as the *hypervolume* of that thing.

But definitions are like babies: giving them a name is one thing, but in order to handle them properly you do need something more than that.[112] In this case, we need formulas.

For instance, how about the hypervolume of a hypercube in n dimensions whose edges all have length equal to 1? As you may have guessed — observe, find pattern, generalise, pat yourself on the back and have a glass of cask strength port barrel aged single malt whisky to celebrate — you get $C_n = 1 \times 1 \times \ldots \times 1$, where the number 1 is to be repeated n times. Not that it matters though, the result is just 1. Technically speaking, this is no longer expressed in terms of square metres or cubic metres, but rather the n-dimensional version thereof. Something like 'hyper-n-metres' — there is plenty of work for people inventing neologisms in hyperspace.

That was relatively easy, right? Looking at these values for C_2 and C_3, which were both equal to 1, it makes sense to assume that this pattern will repeat itself. If someone asks you to predict how the sequence 1, 1, . . . will continue, 'Another number 1?' is the most obvious guess (now that I think of it, 42 is not the answer to life, the universe and everything after all then). People call this an educated guess, but it turns out you can prove that it is not just that: it is the actual value, which means that $C_n = 1$ for all possible values for the dimension n. Visualing hyperboxes may be a pain in the derrière, but at least we know how much goes into them.

8.2 Facultative material

Things get slightly more complicated in the case of hyperballs though. Even trying to *guess* the value B_n becomes a problem here, because it boils down to guessing how the sequence that starts with $B_2 = \frac{\pi}{4}$ and $B_3 = \frac{\pi}{6}$ will continue. Is it $\frac{\pi}{8}$? Or rather $\frac{\pi}{9}$? It

---

112 Other similarities are: people have a natural tendency to find their own definition the prettiest, and certain people may call your behaviour 'rather erratic' when you keep them in the freezer at night.

turns out that it is way more complicated than that: you have more chances to see an owl with a candle on its head being swallowed by a black hole, than to guess what the next number in that sequence will be.

'Wait, let me try!'

Okay, go ahead, I will wait outside until you're tired. So, time to jot down a few guesses ... What do you think will be the value for B_4? Unsurprisingly, there is a formula for B_n which works for all dimensions *n*. If you plug in the value *n* = 4 it leads to the following answer:

$$\frac{\pi}{4}, \frac{\pi}{6}, \quad \text{<insert drum roll>} \quad \frac{\pi^2}{32}, \dots$$

So as you can see from the expression for B_4 above, not only does the number 32 pop up here, also the number π is suddenly supposed to be squared. You probably did not see that coming, did you? In sharp contrast to that owl from the previous paragraph. 'But owls are nocturnal creatures, so it would have been too dark to see it being swallowed by a black hole anyway!' Well, that is not completely true, as I did mention it had a candle on its head.

Now suppose you want to know how the sequence continues with B_5, B_6 and so on. 'That's all right, I can handle that: just give me the formula for B_n and I will plug in the dimension *n* myself.' I really appreciate your enthusiasm, but I am afraid it's not that easy. 'How do you mean? We're almost 200 pages deep into this book, I can handle another formula by now.' Well, okay, here we go: the hyper-volume B_n of a hyperball in *n* dimensions is given by the following number:

$$B_n = \frac{1}{\Gamma(1+\frac{n}{2})} \left(\frac{\sqrt{\pi}}{2}\right)^n$$

<takes a calculator, starts tapping>

'Wait. What?'

Exactly my point, yes. It looks like a fairly standard formula at first sight, doesn't it? It contains the number 2, an exponent *n*, a square root and the number π (so according to Murphi's Law, it cannot be wrong). But the thing that probably stopped you in your tracks is the denominator: how do you feed that symbol Γ into your calculator? Apart from the obvious numerical buttons and the standard operations (+ / − / × / ÷), most machines also come with more advanced options like a built-in cos button, or a sqrt button[113]. However, they rarely come with a Gamma button, so you cannot even use the formula in the box because it is not clear what this Gamma function does (that's how you pronounce the name of this function), nor how you can feed it into a calculator.

The reason for that is the following: this mysterious Gamma function is a standard example of a so-called special function, which belongs to the class of non-elementary functions. Their counterparts are — woop-woop, that's the sound of a no-brainer — the elementary functions, which are the functions that you have surely met in your high school maths textbooks: the power functions, the square and higher order root functions, the exponential functions, the logarithms and trigonometric functions — together with their inverses and a few hyperbolic nephews (creatures like arcsin and cosh). So, an expression like

$$\sin\left(\sqrt[3]{\frac{2x^2 + \sinh\log_3 x}{\ln(x - \sqrt{x+6})}} + 2^{\tan 6x + \arctan \sqrt[5]{\cos x}} \right)$$

---

113   This is not a euphemism for a part of the female reproductive organ by the way, it's the button you press when you want to calculate the square root of something.

looks about as pupil friendly as a lens bottle containing tabasco, but it is still elementary. However, it turns out that not everything can be described in terms of these functions: somewhere along the way, more advanced functions turned up and mathematicians have dubbed these 'special functions'. 'How do you mean, they turned up? How come I've never met them then?' Well, it all depends on the kind of problems you had to solve at school of course, but in most cases these elementary functions are powerful enough to tackle them. Not only that, they can handle a fair amount of being manipulated, which comes in very handy when you have to do calculations. If you start from elementary functions and perform some simple mathemagic (tricks like adding and multiplying functions, or even taking derivatives), you will end end up with a member of the Elementary Function Family again. And this is a good thing, yes. Just think of the resistance against inheritance tax: people like it so much better when everything stays in the family.

But sometimes these elementary functions cannot help you: every once in a while you may bump into a problem whose solution cannot be captured in the language of these 'simple' functions. Think of it this way: you can make a lot of stuff with the standard Lego bricks, but in order to build the Millennium Falcon from the *Star Wars* Saga, you'll need a special expansion set. A typical situation in which this can happen — elementary functions not being powerful enough to capture the solution — is in calculating integrals (the horror). For instance, suppose we start with two standard Lego bricks: the function sin(x) and the function x (see the graph below).

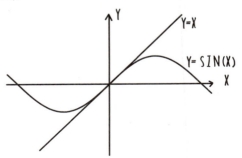

You can add these bricks or you can multiply them: the result will still be an elementary function. 'And what if I divide them?' Same story, it is still elementary. Provided you remember how to do this, you can even take the derivative of all these combinations: it will *still* be elementary. Alas, things seem to change drastically when you start integrating these simple functions and their combinations. It will still work in some cases, because a relation like

$$\int_0^x \cos(t)dt = \sin(x)$$

proves that the integral of an elementary function can again be elementary (pfew!), but it does not work in general (yikes!). Consider, for instance, the following integral, which again contains nothing but harmless elementary functions:

$$\int_0^x \frac{\sin(t)}{t} \, dt \; .$$

Not only is this a nasty problem in itself — you wouldn't be the only one who'd prefer swallowing a jalapeño to solving an integral — it's a *hard* nasty problem, because you can't solve it using the functions you've learnt at school. Despite the fact that you formulated the problem in terms of standard Lego bricks, you will need an expansion set to solve it: the complexity of the solution does not match the simplicity of the question. That's like putting a meat pie into the oven, and taking a banana soufflé out of it: not something you'd expect to happen.

Whenever a situation like this happens to mathematicians — bumping into a problem or an object that doesn't seem to fit into any of the existing boxes — they almost seem to honour it, giving the unexpected result its very own name and upgrading it to a new topic of interest. This is like the difference between a doctor saying that she has absolutely no idea what you are suffering from, and a doctor using a polysyllabic scientific term to describe your condition: in both cases you are not one step closer to recovery, but the latter seems to be a breakthrough as it exudes an air of professionalism.

The integral above forms no exception to this rule: perhaps not surprisingly, mathematicians called it the 'sine integral function'. In a sense, it is just another button missing on the scientific calculator: Si(x).

In the Special Function Zoo, you can find all sorts of strange creatures which got their own name because they didn't seem to fit into the existing framework: there is the error function, the Theta functions, the Hahn-Exton q-Bessel functions, the associated Legendre functions, the hypergeometric functions and so on. One of the better-known special functions is in fact the Gamma function, which appears in our formula for the hypervolume B_n. And just like the sine integral function Si(x) from above, it is also defined in terms of a nasty-looking integral involving nothing but elementary functions:

$$\Gamma(x) = \int_0^\infty t^{x-1} e^{-t} dt$$

So once again, because people kept bumping into this particular integral — it appears in all kinds of problems, including fluid dynamics, quantum physics, number theory, statistics and writing a chapter entitled 'Too Much Space' in a book on mathematics — but had absolutely no idea how to calculate it, mathematicians just gave it a new name and started writing papers about this new member of the family of special functions. When it comes to getting to know a function, there are several techniques: you can write a computer program which approximates the value in a bunch of points and tabulates these, you can prove mathematical formulas relating it to other functions (preferably ones that you already know), or you can take it out on a romantic date.

Speaking of which: what you can also do is plot the graph of the function. This then allows you to read off the value as a number on the Y-axis. The picture below is the graph of the Gamma function (for positive values):

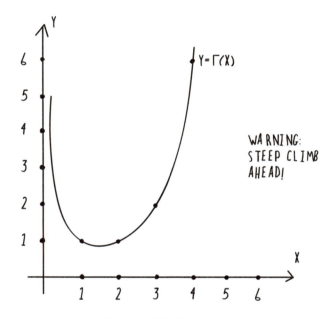

The graph of the Gamma function

x	1	2	3	4	5	· · ·	24	· · ·
Γ(x)	1	1	2	6	24	· · ·	25,852,016,738,884,976, 640,000	· · ·

If you look at this table with values for the Gamma function, one thing is hard to miss: it seems to increase rapidly. Plugging in the value $x = 24$ already gives you a huge number in return, so it's fast-growing to say the least. But there is another thing to be noticed here: although the numbers 1, 1, 2, 6, 24 may look arbitrary to you, there is a simple pattern behind them. Starting from the first number, it seems like the rest is found by doing times 1, times 2, times 3, times 4 et cetera. So the next entry would be 120.

This essentially means that for natural numbers $k \in \mathbb{N}$, we have that $\Gamma(1+k) = 1 \times 2 \times 3 \times 4 \times \ldots \times k$, which can also be written as the number $k!$ (this number should be read as 'k factorial', not as a loud K).

Putting both observations together, we may thus conclude that adding an exclamation mark to a positive integer number $n \in \mathbb{N}$ is actually a nice trick to make big numbers. If you thought Googolplex was a lot, wait until you meet Googolplex factorial — that's just gross.[114] For instance: if you have a deck of 52 cards, then the total number of possible arrangements is equal to 52!. Indeed, there are 52 choices for the first card, 51 choices for the second card, 50 choices for the third card and so on. The total number is thus equal to $52 \times 51 \times 50 \times \ldots \times 2 \times 1$. You can feed this number into your calculator, it will either give you the digital version of a middle finger (Error), or it will spew out the number 8.0658E+67, which is more than the number 8 followed by 67 zeroes (or 10^{67}). To give you an idea: the number of stars in our visible universe is estimated to be 10^{24}, and there are approximately 10^{18} grains of sand on our planet (rough calculations by a group of scientists at the University of Hawaii — a scientist's life can be a beach). I guess natural numbers are like text messages: go easy on the exclamation marks, because they can really blow things out of proportion.

---

114   But not gross enough to beat the mathematicians using Donald Knuth's up-arrow notation in the 'name the bigger number competition' from chapter 3. Sorry to burst that bubble. But that's what arrows do to bubbles.

Definition 54.

Factorial one: *there is an important difference between 'one facto-rial' and 'factorial one'. The former is a number (equal to 1! = 1), whereas the latter is a vital indicator of someone's emotion. You are probably familiar with this phenomenon of people adding so many exclamation marks to their message that it undermines the actual contents (if not: just check your local newspaper readers' forum). A comment like 'OMG, that dude was totally offside !!!!!!!!!!!' is meant to convey the message that a certain player's goal should have been cancelled by the referee, but I usually find myself focusing on the number of exclamation marks (let alone the wrong usage of the space bar after the final word).*

Exclamation marks	Conveyed Message
!	you made your point
!! or !!!	slightly disproportionate (but I guess it *really* affected you)
!!!! to !!!!!!	you are exaggerating now (are you emotionally unstable?)
!!!!!!! and beyond	what was your point again?

What really puts me off though, is a comment like 'OMG, that dude was totally offside !!!!!!!!!!1': it seems I have a hard time sympathising with people who have poor shift key finger muscle control (besides, how can you get so emotionally upset about soccer in the first place, it's just a game between two teams of !! players). Oddly enough, adding factorials in mathematics does not have the same effect: 6! = 720, so you might think that 6!! would be a lot more. On the contrary, the double factorial stands for a leapfrog version of the normal one: 6!! = 6 × 4 × 2 = 48 (the numbers 5 and 3 *were omitted). But if you really insist on being factorially dramatic, you can always throw in a few brackets. Numbers like $(((((6!)!)!)!)!$ are huge, but I will always think of them as the result of a cat kneading a keyboard.*

There is a saying that if it looks like a duck, swims like a duck and quacks like a duck, then it probably is a duck. In a sense, this applies to the Gamma function too: we have seen that although the Gamma function is defined through a technical, complicated formula — as an integral, straight from the depths of hell — it is actually very easy to calculate the number $\Gamma(k + 1)$ when $k \in \mathbb{N}$ is a natural number, because it then simply reduces to the factorial. So you might argue: 'If the Gamma function looks, swims and quacks like a factorial, doesn't that mean that it simply *is* the factorial?'

Well, not exactly: whereas the factorial can *only* be used for integers in $\mathbb{N}$, the Gamma function can *also* be used for real numbers in $\mathbb{R}$. If the former is a penknife, then the latter is a Swiss Army knife: it is a bit overkill when it comes to peeling apples, but indispensable if you are locked in someone's garage and you need to build a cannon from two oil drums, PVC pipes and 12 boxes of canned peaches. So yes, the number $\Gamma(53)$ is just 52!, which means that there is absolutely no need for the integral in that case. But a number like $\Gamma(1 + 3\pi)$ can no longer be thought of as the number of ways to shuffle a deck of cards — well, unless you do know how to shuffle 3π cards — so in order to know this value, you will have to stick to the integral expression. All in all, the Gamma function is the factorial exclamation mark for non-natural numbers.

'Gamma(OMG, this is so cool + 1)'

Okay, it's finally time to get back to our original problem, which was finding the hypervolume B_n of a hyperball in n dimensions. I already gave you the formula, and in the meantime you know what the Gamma button does, so we will calculate this number B_n for some dimensions n.

n	2	3	4	6	8	$\cdots$	100	$\cdots$
B_n	0.7853	0.5235	0.3084	0.0807	0.0159	$\cdots$	10^{-70}	$\cdots$

What you can see here is that the hypervolume of a hyperball decreases as the dimension n increases. If this seems normal to you — God knows what happens in higher dimensions — recall that the hypervolume of the hypercube was *always* equal to 1, regardless of the dimension in which it was sitting. And even though I cannot imagine such a hypercube, this does not feel counterintuitive to me: you can keep extending the thing, pulling at it in new directions, but always over exactly one unit of length. So you are going from length over area to volume and beyond, constantly adding space (under the guise of extra dimensions), but one thing remains invariant: C_n is equal to 1. In some sense, this reminds me of Indian cuisine: if you throw in some spices, you can add depth to the dish and elevate its flavour profile — thereby giving the tasting experience a whole new dimension — but in the end it's still one unit of delicious curry.

Oddly enough, the volume of the hyperball doesn't have this 'invariance property': you can still add space — turning a circle into a sphere and so on — but the more you do this, the smaller B_n becomes. Even though we're not messing around with the radius of the hyperball: the wombat's leash is always exactly half of the hypercube's side length. And yet, the size of the hyperball shrivels into absolute nothingness as the dimension grows.

This conclusion is a rather mind-boggling one: despite the fact that the hypervolume of the box remains constant ($C_n = 1$), the hypervolume of the maximal hypersphere that fits into that box will disappear ($B_n \approx 0$). What this essentially tells you, is that 'all the space' in a hypercube is to be found in one of its many corners, whereas there is nothing to be found 'in the central area'. This is exactly how I remember teenage dance parties *before* 11pm: an empty dancefloor, but plenty of predators standing against the wall — eyeing the prey which had to be caught before our parents picked us up later that night (yet another kissing problem). Like I said earlier, *after* 11pm it was mostly moving around and rotating — trying to keep our growing manhood under control.

As you may have guessed, this makes living in high-dimensional spaces rather uncomfortable. When you add it all up, for instance, I think that some people spend less time looking for a partner than for their keys (even when they *know* they're in their handbag). This doesn't really get better when the number of dimensions gets bigger: not only does the size of your coffee table shrink, leaving them in a corner doesn't make sense either. Most living rooms have eight corners — four of which are rarely known to harbour lost objects, unless you take stray speakers into account — a room in 287 dimensions (which is still a long way from infinity by the way) already has *more* corners than electrons in our observable universe. Then again, who needs keys anyway? It is relatively safe to leave the door unlocked in 287 dimensions, since your house has even more walls than corners, so it does take a stubborn burglar to find your door. And although it's quite unlikely that you will catch your children playing soccer indoors, for the simple reason that the size of their ball will shrink into nothingness, I do not recommend you put them in a naughty corner: it might take them more than a few decades to be back for supper.

Definition 55.

Handbag: *a fashionable container, often made of fabric or leather, mostly carried around by the female species. May contain items such as eye-liner and lipstick, a purse, a mobile phone, handkerchiefs, keys, cookies, fashion magazines, pets, lunchboxes, cameras, shoes, laptop computers, perfume and deodorant, glasses, plastic bags, a bottle of sparkling water, contact lenses, several pieces of fruit, cutlery, prescription drugs, a spare handbag, chewing gum, an inflatable cushion, books, peanuts, elastic hair bands, sunscreen, lip balm, condoms, pocket dictionaries, peppermints, a toothbrush, a comb, shampoo samples, a pair of socks, gloves, et cetera.*

The average weight and street value of a handbag is estimated to be the equivalent of 10 kilograms of cocaine.

Statisticians, computer scientists and discrete mathematicians[115] are very well aware of this nuisance of increased dimensionality, as they encounter its consequences in their research. They therefore refer to it as 'the curse of dimensionality' or 'the combinatorial explosion'. These words may have no meaning to you, but just like hippopotomonstrosesquipedaliophobia and aquagenic urticaria, you don't need to be a qualified doctor or psychologist to realise that it has nothing to do with giant fluffy rabbits wearing chocolate helmets, cuddling you to sleep. It not only sounds bad for you, it is bad for you. Let me give you an example of this principle: suppose you want to cook dinner for your partner, and you've got a list of things that he or she likes or dislikes.

Likes	Dislikes
green asparagus	beetroot burgers
goat's cheese	spinach and pomegranate salad
crispy bacon strips	spicy fried chicken
winegums	celery stalks
blueberry muffins	charred broccoli
charred Brussels sprouts	buttermilk pancakes
grapefruit	beer-battered onion rings

---

115   Discrete mathematics is a branch of mathematics in which one studies structures which are discrete rather than continuous. It should not be confused with discreet mathematics, although not much is known about this branch: researchers rarely seem to publish their results.

You study the list, and you decide to prepare the following dishes: rack of lamb with broccoli cheese, pomegranate rice and carrot cake with buttermilk cream. 'Why would I do that, it has all the wrong ingredients?' Well, as is often the case in mathematics, it all depends on the definition you're using. What do you call 'wrong' here? Because based on the list, it is very clear that your partner loves dishes and ingredients that contain an *even number* of r's. The menu that I suggested meets that criterion perfectly, so why wouldn't you? The problem here is that by considering letters, you are working in a space with 26 dimensions: one way to look at this list is to identify each entry with a point in this space. We could do this in terms of coordinates, so we would then, for instance, have that grapefruit corresponds to the point

$$(1, 0, 0, 0, 1, 1, 1, 0, 1, 0, 0, 0, 0, 0, 0, 1, 0, 2, 0, 1, 1, 0, 0, 0, 0, 0).$$

Each number hereby indicates how many times a letter occurs (from A all the way to Z). If you do this, you end up with a collection of 14 points (the list) in a 26-dimensional space: this is literally too much space, which means that you will always be able to find a criterion that separates certain points from other points (the likes and the dislikes in our case). Put differently, if you really want to stick to letters in order to decide what you will be cooking for dinner, you need w*ay* more data points (entries in the list).

What I just described here is one of the main problems in the world of Machine Learning and Big Data — two current topics in computer science that buzz harder than an electric bumblebee ringing the doorbell. While the number of parameters (dimensions) increases linearly, the number of data you will need to make 'relevant' conclusions increases exponentially. And like we already learnt in chapter 6 when we compared **P** against **NP**, unless we are talking about your salary or the number of holidays, exponential increase is something we rather prefer to avoid.

8.3 Rounding things off

In the end, I suppose we can safely conclude that life in three dimensions is not so bad after all. It may get a bit crowded at times,[116] but at least we don't have to be afraid of getting lost in one of its disconcertingly many corners. Some physicists claim that life as we know it would not even exist in a universe which does not have exactly three spatial dimensions and one temporal dimension. Yes, we are excluding the extra dimensions predicted by string theorists here, but these are conjectured to be rolled up on a nanoscopic scale, so they don't really matter anyway.[117] For instance, gravity would be too strong for solar systems to form in two dimensions, and too weak in four dimensions. No solar systems. It's hard to grasp how boring that would be: no need for a funky creation myth, no pizzas to be sliced and no coconuts having a bad hair day. Explanations like these — involving gravity or other physical laws and constants which, as by some sort of miracle, seem to be *exactly* right for the formation of life — are to be filed under the so-called anthropic principle. This is a philosophical argument based (to a greater or lesser extent) on the idea that the universe is as it is because we, humans, are here to observe it that way. This is a bit like gazing at your navel, and concluding that your belly button is the way it is because you've got eyes to look at it — but then on a deeper level.

Even more profound is the question whether a universe without planets and humans to inhabit them still comes with mathematics. For most people, a void cosmos in which maths still exists would be like stepping on a Lego brick while being chased by a zombie — a nightmare within a nightmare — but for some this question, and the hunt for an answer to it, is precisely what rocks their world. I alluded to the connection between maths and philosophy before, when I talked about the differences between, for instance,

---

116 See also: Hong Kong.

117 This makes them very different from maki sushi and sleeping cats: rolled up too, but hard to imagine a meaningful life without them.

Platonism and Intuitionism, but I think this is the ultimate question: is mathematics something that transcends the reality of a universe, or is it woven into the fabric of space-time and did it come into 'being' when the universe began (which would mean that different universes come with different sorts of mathematics)? Some people even believe that, on its most fundamental level, the universe is nothing *but* mathematics. Although these are deep and intriguing questions, I have to admit that when I am lying on my back — pensively gazing at the stars on a dark summer night — I mostly find my brain focusing on even more thought-provoking mysteries, like 'does it even make sense to translate "Do not touch" into braille?'

In any case, it seems like the conditions of that cosy little corner of the universe in which you were reading this book were just right for you to reach its final pages. We have been on a wonderful voyage together, through the quirky universe of mathematics — that fascinating space in which you hopefully roamed around with the same enthusiasm with which I always find myself drifting around Tokyo: fascinated by its language, trying to make sense of things, impressed by both the overall structure and the finer details, no longer afraid to get your teeth into all the delicacies it has to offer.

Recommended listening

Artist	Song title
Scooter	Hyper, Hyper
Linkin Park	In The End
Year of No Light	Disorder
Modern Life is War	Find a way
Amorphis	Too Much To See
Zeal and Ardor	We Can't Be Found
Grails	A Mansion Has Many Rooms
Life of Agony	Lost at 22
Joy Division	Isolation
Upcdownc	Z-more

9

Space For Improvement

> *I am just a child who has never grown up.*
> *I still keep asking these 'how' and 'why' questions.*
> *Occasionally, I find an answer.*
>
> (Stephen Hawking)

As you have now reached the final pages of this book — well done, your Perseverance Points will be delivered shortly — I think it is only fair that I now offer you a chance to test yourself, to see how much you have learnt to think like a mathematician. Answers are provided at the end of the quiz!

Question 1: *What is the most boring integer number?*

(a)   37

(b)   there is no such thing as the most boring integer

(c) π

Question 2: *Suppose someone asks you for a random prime number, what do you say?*

(a)   2

(b)   57

(c) $2^{82,589,933} - 1$

Question 3: *Who would you rather have been?*

(a)   Leonhard Euler

(b)   Paul Erdős

(c)   Evariste Galois

Question 4: *Which of the following do you consider to be the most beautiful mathematical formula?*

(a) $e^{i\pi} + 1 = 0$

(b) $\dfrac{1}{\pi} = \dfrac{2\sqrt{2}}{9801} \sum\limits_{k=0}^{\infty} \dfrac{(4k)!(1103 + 26390k)}{(k!)^4 396^{4k}}$

(c) $\Big(\text{food} - (\text{taste}) \times (\text{colour}) \times (\text{smell}) \Big) \times \text{sarcasm} = \text{dark tofu}$

Question 5: *What is the next number in the following sequence?*
$$2 , 4 , 8 , 16 , \ldots$$

(a)   32

(b)   31

(c)   any number will do

Question 6: *If you were a mathematician and you had a dog, what would you call it?*

(a)   Cauchy

(b)   Max

(c)   Ket

Question 7: *Which of the following does not exist?*

(a)   the Cox-Zucker machine

(b)   the Alpher-Bethe-Gamow paper

(c)   the Ham Sandwich Theorem

Question 8: *Can you name the following bands?*

(a) $A \div D$

(b) $m \in M$

(c) $(x - y)(x + y) + y^2$

Question 9: *Which of the following answers is correct?*

(a) $\sqrt{2\pi}$

(b) $\dfrac{211 + \sqrt[13]{619}}{17}$

(c) 42

Question 10: *Which disjunction type is used in the sentences below?*

(a)   Are you allergic to mayonnaise or peanuts?

(b)   The first thing I always do upon arrival in Tokyo is to eat a bowl of ramen or a plate of tempura.

(c)   Should I stay or should I go now?

Question 11: *Which of the following descriptions is correct?*
Bonus question: *Where to insert the tumbleweed?*

(a)   Log function: to be put on a campfire

(b)   Sine function: to make yourself known

(c)   Cosine function: to make yourself known too

(d)   Square root function: to make a box of vegetable soup

(e)   Nilpotent: weak

(f)   Idempotent: as strong as the other

(g)   Equipotent: as strong as a horse

(h)   Vermillion: bloody many

(i)   Diophantine equation: comparison of two elephants

(j)   Divergent: male PADI-holder

(k) Minion: a lot of ϵ's together

(l) Global maximum: best orgasm ever

(m) Local maximum: a quickie

(n) 2^7: De La Soul's answering machine

(o) Power of 3: a forest

(p) Power of 8: so full I can't even *think* of food now

(q) Power of 9: a harsh German negation

(r) Power of 42: answering all questions

(s) Decomposable group: recyclables which can be added and inverted

(t) Sporadic group: a bunch of mushrooms

(u) Horizontal axis: guillotine

(v) Polar coordinates: to describe positions below zero

(w) Triple integral: a complete strong pale ale

(x) Manifold: the opposite of a crumpler

(y) Saddle point: a horse's back

(z) Knot theory: practice

Answer 1:

(a) When people are asked to say a random number, it turns out that 37 is the most common answer. Well, it all depends on who's asking that question of course: if Rachel Riley[118] were to ask me for a number, I would give her a nine digit one — my phone number. The number 37 actually has remarkable properties. For instance, if a, b and c are three integers such that abc is a multiple of 37, then this holds for bca and cab as well. For example, $629 = 17 \times 37$, $296 = 8 \times 37$ and the last combination gives $962 = 26 \times 37$. Also, any number of the form $abczyx$ (where the first three digits are increasing and the last three digits are decreasing) is a multiple of 37. For example, one has $234{,}876 = 6348 \times 37$. It is also a lazy caterer number: with eight cuts, you can get as many as 37 pieces.

I know this probably sounds like too much of a coincidence, but I wrote these words the day before my 37th birthday. Not necessarily a memorable age, in my opinion, although it was the first time someone sent me one of those cards which say 'You are like wine: you improve with age'. I sincerely hope that is indeed the reason, and not 'You should be kept in a dark cellar for a few years, until one Christmas Eve you will be sniffed at before being allowed at the dinner table.' In any case, this earns you 2 points.

(b) This answer gives you 3 points. As a matter of fact, you can even prove why this is true: suppose there was such a thing as 'boring numbers'. If we were to order all the boring numbers from small to big, the first boring number would have a rather remarkable property: it would be the very *first boring number*, which by definition means that it would be a very interesting number (it would be the unique number with this property).

---

118 An English television presenter and mathematician with a, erhm, beautiful mind.

From which we conclude that there simply can't be such a thing as a boring number.

(c) Really? Come on, you can do better than that: it is not even an integer. You get $\sqrt{-1}$ points for this. Unless you are the adjudicator from chapter 2, and you finally finished counting.

Answer 2:

(a) You get a point for this, since 2 is definitely the oddest possible answer. When it comes to puns in mathematics, this may be an old chestnut, but you have to admit: it is a prime example. Well, I am going off on a tangent here, but there is a pun about derivatives too.

(b) 'But wait a moment, $57 = 3 \times 19$ is not even a prime number.' True, and yet this particular number is known as the *Grothendieck prime*, named after the German-born French mathematician Alexander Grothendieck (1928-2014). He is nothing less than a saint for those working in the field of Algebraic Geometry, a term which probably sounds to most people like 'Mayonnaise Spongecake' sounds to me: a concoction of two things that even separately bring me out in a rash, let alone combined.

According to legend, someone in a mathematical conversation with Grothendieck suggested that they should consider a particular prime number as an example. 'Do you mean an actual number?' Grothendieck asked. The other person replied, 'Yes, an actual prime number,' to which Grothendieck answered, 'All right, take 57 then.'

He was one of the greatest mathematicians of his era, and the list of things named after him is quite long. It contains objects such as Grothendieck's theorem, the Grothendieck category, Grothendieck's spectral sequence, Grothendieck's trace formula, a Grothendieck space, the Grothendieck topology, the Grothendieck group, the Grothendieck connection, Grothendieck's constant and so on. I am actually surprised you cannot order a Grothendieck Latte at Starbucks: 57 centilitres of French roast in a beer stein.

(c) As it stands now, this is the largest known prime number (found in December 2018), consisting of 24,862,048 digits. This means it would require something like 8,000 additional pages if I were to include it in this book. If you haven't got time to read that, no worries: there is this Indian Guinness World Record holder who might do the audio version of the book. You get 2 points for this.

Answer 3:

(a) You have chosen one of the most prolific mathematicians of all time. Leonhard Euler was a Swiss mathematician, physicist, astronomer, logician and engineer (1707-1783), and his collected works fill enough volumes to make even the best mathematicians feel like they have done absolute diddly-squat. I once read in an algebra textbook that 'doing research is like jumping on a fast-moving train'. I suppose Euler was lucky in that respect, he was already there when they were still drawing the plans for the platform. He was also very lucky not to be Belgian: he could have been known for a completely different list of jobs, because he got tired of waiting for that damned train — which isn't even that fast. You get 1 point.

(b) You have chosen one of the most prolific mathematicians of the 20th century, known for his social approach towards maths: Erdős is known for having collaborated with more than 500 people over the course of his career. You get 2 points.

(c) You have chosen the main character in one of the most romantic stories about mathematics and its heroes. This

French prodigy died at a young age (1811-1832) from wounds suffered in a duel, the true motives for which are still a matter of debate. Some say it has to do with a broken love affair with a woman, others believe the incident was stage-managed for political reasons. No matter the cause, Galois was so convinced of his impending death that he pulled an all-nighter, literally the last hours before his opponent pulled a gun, writing what would become his mathematical testament — a famous letter outlining his ideas, and three attached manuscripts. Given the fact that most people would rather do something else when they know it's their last night — finally opening that cask strength port barrel aged single malt whisky for which special occasions never seem to be special enough, for instance — this earns you as many points as you want.

Answer 4:

(a) If this is your answer, you are definitely not alone. As a matter of fact, a *Mathematical Intelligencer* poll voted this formula as the most beautiful formula in mathematics (1988). It is known as Euler's identity, and it is nothing short of a Mathematical Dream Team: it combines two essential integers (1 and 0), two fascinating real numbers which show up everywhere in nature (π and e) and the mind-boggling square root of -1 (the complex unit i, which satisfies $i^2 = -1$). You get 1 point.

(b) As you may have guessed in the meantime, this formula for π is thanks to the Indian mathematician Ramanujan (see chapter 2). If this one was also whispered into his ears by Hindu goddess Namakkal during his sleep, I will stop reading books and start working on my dreaming techniques. Or as Martin 'Lucid' King once said: *I will have a dream.* You get 2 points.

(c) Ah, it seems like you prefer the kind of mathematics which involves more letters than numbers: this earns you n points with $2 < n \leq 5$.

Answer 5:

(a) This is the most obvious answer, which is what most people will say: it's where the bus drivers from chapter 3 will sleep. No big deal.

(b) Actually, there is a clear-cut way to see why the answer should be 31. If you put two dots on a circle (the circumference, I mean) and connect them, you get two pieces (circular segments). If you put three dots on a circle and connect them, you get four pieces (a triangle and three circular segments). If you start from four points, you end up with eight pieces (four triangles and four segments). Similarly, five dots lead to 16 pieces (11 triangles and five segments). So far so good, but if you start from six dots on a circle you can get at most 31 pieces. I say *at most*, because it can be 30: you lose a triangle in the center if you start from a regular hexagon (you'd miss region 18 in the picture below). So there you go, less can be more in maths too.

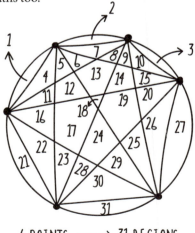

6 POINTS $\longrightarrow$ 31 REGIONS

(c) It is a common frustration amongst mathematicians that in IQ tests, one is always expected to give 'the most obvious' answer to questions of the form 'what is the next number in the sequence?'. As a matter of fact, there is a mathematically sound way to justify why *any* number could be the next one. So as long as you don't have a chance to add your explanation, the only thing such a question accomplishes is testing how much you think like an average person.

Apart from that mathematical 'trick' which you can use to make any number fit a given sequence, there are surprisingly many answers for which there is an explanation. You can look this up yourself, using the Online Encyclopedia of Integer Sequences (see https://oeis.org). If you then type 1, 2, 4, 8, 16 into the search box, you get a whopping 720 possibilities to continue the sequence (to be more precise, sequences in which these five numbers appear as a subsequence). Not only several different explanations for why the next number could be 32, but also many other solutions. For instance, it could be

$$1, 2, 4, 8, 16, 77, 145, 668, 1345, 6677, 13444, 55778, \ldots$$

This is the so-called RATS-sequence: Reverse, Add and Then Sort. We reverse the digits in 16 (which gives 61), add it to 16 and then sort the digits: this gives 77.

Answer 6:

(a) The French mathematician Augustin-Louis Cauchy (1789-1857) is (among many other things) known for his residue theorem. This is a crucial result in complex analysis, which says that 'residues can only be found at poles'. Obviously, this is the top answer (3 points).

(b) Max is both a mathematical notion and a dog's name, but they are both fairly standard: so this one earns you 2 points.

(c) Ket may qualify well as a dog name — it has only three letters and it sounds okay when called out loud — but it is the most risky option: as it is an abbreviation for a recreational drug, shouting your dog's name in public may easily be confused with soliciting for illegal substances. On top of that, explaining your loyal friend's handle to the police officer won't really help you: it would not be the first time that the words *'no worries, a ket is nothing but a vector in the Dirac formalism for quantum mechanics, and together with a corresponding bra vector it gives a quantum state'* could be interpreted as a proof that you've already licked the toad's back. For the sake of your safety: 1 point.

Answer 7:

Sorry, this was a trick question: they all exist!

(a) The Cox-Zucker machine, named after David Cox and Steven Zucker, is an algorithm used in algebraic geometry which first appeared in their 1979 paper 'Intersection numbers of sections of elliptic surfaces'. I could try to explain what it does — using words such as 'projective varieties', 'Mordell-Weil groups' and 'elliptic surfaces' — but I don't want to be foul-mouthed.

(b) This refers to a paper in cosmology from 1948, created by Ralph Alpher (still a PhD student at that time) and his advisor George Gamow. The latter decided to add the name of his friend Hans Bethe to the list of authors, in order to create a play on the first three letters of the Greek alphabet. The former was somewhat dismayed by this, as he was afraid that the inclusion of another eminent physicist on the list of authors would overshadow his own contribution to this work, but he couldn't stop his advisor's whimsy. In sharp contrast to Andy Slurping, the more stubborn colleague of Cox and Zucker.

(c) This theorem is a result in *measure theory*, a mathematical discipline that I briefly touched upon when I wrote about unmeasurable sets (see the Banach-Tarski paradox). It says that whenever you have three objects, arbitrarily shaped and representing a chunk of ham and two loafs of bread, you can always simultaneously bisect these objects using a single cut (with a two-dimensional plane). It should not be confused with the sandwich rule, which says that at a reception carnivores will always eat the cheese sandwiches first, before they turn their attention to the ham sandwiches, hereby leaving the vegetarians empty-handed. There is a simpler version as well, called the pancake theorem: two flat objects, also arbitrarily shaped, can be bisected using a single cut (with a one-dimensional line).

Answer 8:

You get 1 point for every correct answer here.

(a) AC/DC

(b) Eminem

(c) The xx

Answer 9:

(a) Murphi's Law earns you 1 point here.

(b) This answer gives you 5 well-earned points. My experience in grading exams has taught me that when students have to solve an equation or calculate an integral, they seem to be perfectly okay with their solution if it belongs to the set:

$$S = \left\{ 0, 1, \sqrt{2}, \frac{\sqrt{3}}{2}, \pi \right\} \ .$$

These numbers seem to exude some sort of confidence, as if they are more likely to be the answer to a question than other 'random' numbers. But when they find their answer to a problem to be, say, 19 or $\frac{7}{3}$, most students will redo the calculation, thinking they must have made a mistake somewhere.

(c) As a rule of thumb: if you can choose 42, just do it — it may earn you 3 points. I once made an exam for undergraduate students bearing a quote in the upper right corner which said '42. *The answer to life, the universe and everything*', and the answer to almost all of the questions was indeed 42. The attentive reader[119] probably noticed that I did say *almost*: the answer to one of the questions was 43. Now if you think *that* was a nasty mindfuck, how about this one: if there is an exception to every rule, how about this one being an exception itself?[120]

---

119 I sometimes wonder whether this mythical creature really exists. Many authors seem to accept its reality, but the attentive reader's alleged abilities are borderline supernatural: he or she seems to have the memory of a herd of elephants coated in USB-sticks, and the speed with which it can predict the author's intentions makes tachyons look like a land snail listening to doom metal.

120 After a quote by James Thurber, which I only found out afterwards.

Answer 10:

If this question didn't make sense, the definition about logical disjunctions (chapter 2) will help you out. So flip back a few pages *or* give yourself 2 points.

Answer 11:

This may catch you by surprise, but this is actually a joke. After all the serious stuff, I thought it would be nice to end this book on a lighter note. You know what, give yourself 1 point every time you had to smile. Add a tiny bonus point if you got the reference to tumbleweed and ϵ.

Bibliography

1. Readers who want to catch up (tomato-based pun level unlocked) on the hot dog story from page 16 can start from the following link: https://www.independent.co.uk/news/world/americas/america-decides-is-the-hot-dog-a-sandwich-or-not-a6726656.html

2. The genuine research paper mentioned on page 23: J. L. Silverberg, M. Bierbaum, J. P. Sethna and I. Cohen, Collective motion of humans in mosh and circle pits at heavy metal concerts, Phys. Rev. Lett. 110.228701 (2013).

3. We gladly thank the Archives of the Mathematisches Forschungsinstitut Oberwolfach for letting us use this picture of Martin Gardner (Author: Konrad Jacobs) on page 26.

4. There is obviously a huge amount of literature on Fermat's celebrated last theorem — on the account of being one of the most (if not the most) successful stories in mathematics — but if you really want to start somewhere I'd suggest reading Simon Singh's book from 1997.

5. Picture on page 56 was taken from https://commons.wikimedia.org/wiki/\File:Euclidian_and_non_euclidian_geometry.png

6. The quote on page 99 comes from John Green's novel The fault in our stars from 2012. It's rather ironic that Green — using the voice of Hazel, the young narrator of the story — makes a mistake in the book when Hazel expresses her lovely sentiments packed in a paragraph containing some facts about infinity. If you ever feel like testing your understanding of the concept of a bijection, this is your chance!

7.   The pioneering paper by Georg Cantor mentioned on page 100: *Über eine Eigenschaft des Inbegriffes aller reellen algebraischen Zahlen*, Journal für die Reine und Angewandte Mathematik **77**, pp. 258-262 (1874).

8.   On page 109 I mentioned the so-called Continuum Hypothesis. The idea behind this hypothesis is fairly simple: there is something like $\aleph_0$ (the cardinality of the natural numbers) and $\aleph_1$ (the cardinality of the real numbers), and we already know that $\aleph_0 < \aleph_1$. But one may wonder whether there is something that lies between these two infinities. Just think of the following example: 3 is less than 4, and you can find something that lies between these numbers. For instance $3 < \pi < 4$. As for infinities, this is a *very* deep question, which has to do with the axioms you start from. The curious reader may find more information on the internet, although it all becomes rather technical at some point.

9.   More on the story about Kit Kat and its middle layer from page 115 can be found at http://www.dailymail.co.uk/femail/article-4092136/Documentary-reveals-Kit-Kats-made.html

10. The proof for the fact that the kissing number in three dimensions is equal to 12 (see the story on page 147) can be found in K. Schütte, K. and B. L. van der Waerden, *Das Problem der dreizehn Kugeln*, Math. Ann. **125**, pp. 325-334 (1953).

11. Ferguson's paper from page 167: *Who solved the secretary problem?*, Statist. Sci. **4**, No. 3, 282-296 (1989).

12. More on the argument of Greg Egan referred to on page 178 can be found at https://johncarlosbaez.wordpress.com/2015/07/20/the-game-of-googol/

13. A news article about the pizza restaurant on the moon (see page 228) can be found at https://www.telegraph.co.uk/news/ science/ space/8734456/Dominos-plans-pizza-on-the-Moon. html

14. John von Neumann's seminal paper about game theory on page 235: *Zur Theorie der Gesellschaftsspiele*, Math. Ann. **100** No. 1, pp. 295-320 (1928).

15. The paper in which it was shown that 4/9 is a magic number related to eating pizza, mentioned on page 173: K. Knauer, P. Micek and T. Ueckerdt, *How to eat 4/9 of a pizza*, Discrete Math. **311**, Issue 16, pp. 1635-1645 (2011). The attentive reader may have noticed that in our examples on page 238, the losing player gets 405/909 of the evil pizza, which is *more* than 4/9. The reason for this is because in their paper, Knauer and co work with pizzas with slices that can have a zero size. Their example can be recovered if you replace all the 1-slices by 0-slices.

16. Readers who want to read more about the strange tilings in more than six dimensions (see page 262) can consult the Wikipedia page on the Keller conjecture.

17. The connection between wombat dung and cubes on page 270 might be something that went over your head (better than on your head). In that case, you may want to check http://www. iflscience.com/plants-and-animals/why-do-wombats-do-cube-shaped-poo/.

18. On page 288, I mentioned the philosophical issues related to maths and the universe. Readers who want to dive into these fascinating theories might want to start with, for instance, Max Tegmark's *Our Mathematical Universe: My Quest for the Ultimate Nature of Reality* (2014) and Paul Davies' *The Mind of God: The Scientific Basis for a Rational World* (1992).